The Science of Climate Migration

This book addresses the nexus between science and migration and examines how the two are inextricably intertwined. *The Science of Climate Migration* primarily addresses the science of global climate change and additionally examines how this change is more than a region being too hot, too cold, too dry, too wet, or too windy; rather it is also about heightened military tensions, political instability, and myriad other factors. History has shown that this change is felt most acutely in developing countries that are least equipped to adapt. This inability to adapt is considered a driver that motivates local residents to find "greener pastures" through migration. Further, the book discusses the increasing need for the implementation and utilization of non-polluting renewables for use in energy production as a means to stave off environmental crises.

FEATURES

- Examines how and why climate change effects and human migration are inextricably intertwined.
- Discusses the increasing need for the implementation of non-polluting renewables for use in energy production as a means to stave off environmental crises.
- Explains how wildlife is also sensitive to shifts in climate and how this in turn affects their migration as well.

The Science of Climate Migration

Frank R. Spellman

CRC Press
Taylor & Francis Group
Boca Raton London New York

CRC Press is an imprint of the
Taylor & Francis Group, an **informa** business

Designed cover images: Shutterstock | Atikan Pornchaiprasit; Shutterstock | Andrey_Popov; Stand of beech trees before subsidence. Artwork by Frank R. Spellman (2020).

First edition published 2023
by CRC Press
6000 Broken Sound Parkway NW, Suite 300, Boca Raton, FL 33487–2742

and by CRC Press
4 Park Square, Milton Park, Abingdon, Oxon, OX14 4RN

CRC Press is an imprint of Taylor & Francis Group, LLC

Library of Congress Cataloging-in-Publication Data
Names: Spellman, Frank R., author.
Title: The science of climate migration / Frank R. Spellman.
Description: First edition. | Boca Raton : CRC Press, 2023. | Includes bibliographical references and index.
Identifiers: LCCN 2022030471 | ISBN 9781032280745 (hbk) | ISBN 9781032280752 (pbk) | ISBN 9781003295211 (ebk)
Subjects: LCSH: Climate change mitigation.
Classification: LCC TD171.75 .S67 2023 | DDC 363.738/74—dc23/eng/20221011
LC record available at https://lccn.loc.gov/2022030471

ISBN: 978-1-032-28074-5 (hbk)
ISBN: 978-1-032-28075-2 (pbk)
ISBN: 978-1-003-29521-1 (ebk)

DOI: 10.1201/9781003295211

Typeset in Times
by Apex CoVantage, LLC

Contents

About the Author

Frank R. Spellman is a retired assistant professor of environmental health at Old Dominion University, Norfolk, VA, and author of over 155 books. Spellman has been cited in more than 400 publications; serves as a professional expert witness and incident/accident investigator for the U.S. Department of Justice and a private law firm; and consults on Homeland Security vulnerability assessments (VAs) for critical infrastructure, including water/wastewater facilities nationwide. Dr. Spellman lectures on sewage and water treatment, homeland security and health and safety topics throughout the country and teaches water/wastewater operator short courses at Virginia Tech (Blacksburg, VA). He holds a BA in public administration; BS in business management; MBA; Master of Science, MS, in environmental engineering; and a PhD in environmental engineering.

Preface

The Science of Climate Migration is the eighth volume in the acclaimed series that includes *The Science of Water, The Science of Air, The Science of Environmental Pollution, The Science of Renewable Energy, The Science of Waste,* and *The Science of Rare Earth Elements: Concepts and Applications* (in production), all of which bring this phenomenally successful series fully into the 21st century. *The Science of Climate Migration* continues the series mantra based on good science and not and "never" on feel-good science.

Science? Migration is about science? Well, it is when it is driven by the elements of science. The science, the driver, is climate change—the push and pull effect. Ask yourself a few "I" questions. If I live in an area where perpetual drought has taken over and a drink of water is costly beyond my means or not available at all, what is my next move? If I can see the air I breathe and it makes me sick, what am I going to do? If my weather and climate abruptly change to the point where it is too warm to do anything, or there is an increase in frequency of storms, including tornadoes, hurricanes, and torrential downpours with accompanying flooding, what am I going to do? If I live along an ocean shoreline and the sea level is rising and flooding me out, what am I going to do? If I live in an area where my livelihood is based on a winter producing a certain amount of annual snowfall and the snow becomes sparse or non-existent, what am I going to do? If I live in a region that was free of pestilence and then suddenly changes and Lyme disease and other vector-transmitted illnesses or worse are occurring, what am I going to do? If I live in a region where suddenly wildlife is mysteriously dying off; birds are leaving; streams and lakes are heated to the point where fish are dying in shoals or schools; and dangerous animals such as alligators, crocodiles, and poisonous snakes and insects are invading and taking up residence, what am I going to do? If I live in an area where crops no longer grow, where out-of-control forest fires are not only common but consistent, what am I going to do? Finally, if climate change affects any aspect of my life whereby it is simply unhealthy or dangerous for me and my family to remain in my area, what am I going to do?

What am I gonna do?

Who am I gonna call? Ghostbusters is a hoax, a fantasy, a fancy, a desire, a vision, but one thing is certain: The Ghostbusters are not available. And, truth be told, we are not dealing with ghosts, presences, apparitions, ghouls, phantoms, or poltergeists . . . no, no way—what we are dealing with is real, authentic, bona fide, and an ongoing issue.

Well, if the conditions stated previously are brought about by climate change affecting you, it might be time to get out of Dodge. To might consider leaving the area and seeking "greener pastures," so to speak. You might choose to migrate. You might choose to hike hundreds or even thousands of miles to find the so-called promised land. And if, once you finally reach the new location and there is some obstacle such as a mountain, desert, river, army of soldiers or human-made walls blocking your entrance, what are you gonna do?

Well, if there are obstacles in your way, you will find a way around or over the obstacle, even if you must breach the wall or whatever—recent history has shown that there is always a way to make the breach.

Anyway, this book is about science tied to migration. It deals with science in two distinct ways. First, it deals with climate change, and there is no doubt about the connection between science and climate change. Second, climate change is about more than it being too hot, too cold, too dry, too wet, too windy, or the ground being too shaky. It is also about heightened military tension and uncertainty. Experience has shown, and it is evident to the nth degree, that change is felt most acutely in developing countries that are least equipped to adapt. This inability to adapt is a "driver" that motivates residents to find those greener pastures. Also, when a select few of the privileged class control whatever assets a region possesses, the down-and-out are out-and-down. It is time to get out of Dodge—by whatever means possible.

Truth be told, this is exactly what is occurring at the present time.

Anyway, about this book. Concern for the environment and for the impact of human-caused or human-exacerbated climate change has brought about the trend (and the need) to shift from the use of and reliance on hydrocarbons to energy-power sources that are pollution neutral or near pollution neutral and renewable. We are beginning to realize that we are responsible for much of the environmental degradation of the past and present—all of which is readily apparent today. Remember, in many ways, there are greener pastures—somewhere. Well, unless the entire globe is contaminated, and climate change is apparent and perceptible everywhere. Moreover, the impact of 200 years of industrialization and surging population growth has far exceeded the future supply of hydrocarbon power sources. So the implementation of renewable energy sources is surging, and along with it there is a corresponding surge in utilization of non-polluting renewables for use in energy production.

Why a text on the science of climate migration? Simply put, studying climate change and its consequences is something we should all do. Many of us have come to realize that a price is paid, forfeited (sometimes a high price) for what is called "the good life." Our consumption and use of the world's resources make us all at least partially responsible for pushing the need to contribute to the climate change environment due to our use of conventional energy sources such as oil and coal. Pollution and its ramifications are one of the inevitable products of the good life we all strive to attain, but obviously pollution is not something caused by any single individual, nor can one individual totally prevent or correct the situation. The common refrain we hear today is to reduce pollution and its harmful effects, everyone must band together as an informed, knowledgeable group and pressure the elected decision makers to manage the problem now and in the future. At this moment in time there is an ongoing push to replace fossil fuels with renewable energy sources—this is where the shift to renewable energy sources comes into play and where the need to mitigate climate change and its causal factors is vital.

Throughout this text, common-sense approaches and practical examples are presented. Again, because this is a science text, I have adhered to scientific principles, models, and observations, but you need not be a scientist to understand the principles and concepts presented. What is needed is an open mind, a love for the challenge of wading through all the information, an ability to decipher problems, and the patience to answer the questions relevant to each topic presented. The text follows a pattern that is nontraditional; that is, the paradigm used here is based on real-world experience, not on theoretical gobbledygook. Real-life situations are woven throughout the fabric of this text and presented in straightforward, plain English to present the facts, knowledge, and information to enable understanding and allow for informed decisions.

Environmental issues related to climate change are attracting ever-increasing attention at all levels. The problems associated with these issues are compounded and made more difficult by the sheer number of factors involved in managing any phase of any problem. Because the issues affect so many areas of society, the dilemma makes us hunt for strategies that solve the problems for all while maintaining a safe environment without excessive regulation and cost—Gordian knots that defy easy solutions.

The preceding statement goes to the heart of why this text is needed. Presently, only a limited number of individuals have sufficient background in the science of climate change and its concepts and applications in the world of industrial and practical functions, purposes, and uses to make informed decisions on 21st-century product production, usage, and associated environmental issues.

As with the other editions in this series, this book is presented in the author's characteristic conversational style. The goal is to communicate with the reader—to do this, one must speak plainly in terms readily understood.

Another point: migration driven by climate change is not only a human phenomenon, but it also drives wildlife to find those greener pastures. We are not alone on this planet; other living things occupy a lot of our space—and rightfully so.

My traditional and ongoing bottom line: Critical to solving real-world environmental problems is for all of us to remember that old saying: we should take nothing but pictures; leave nothing but footprints; kill nothing but time; and sustain ourselves with clean, safe air, water, and soil. If we can abide by all of this, migration to greener pastures may be unnecessary.

Frank R. Spellman
Norfolk, VA

Prologue: Too Many for Too Little

This is not a political book.

This is a science book.

Well, you might or could point out that political science is a science, isn't it?

Whether political science is a science is debatable, dubious, arguable, questionable, and undetermined, for sure.

One thing is certain: Climate and climate change is all about science.

Okay, what kind of science? What science disciplines are involved? What topics need to be studied to become a "climate science person"?

Well, first, when we need to know what is going on with our climate, we need to check with a couple of true experts, specialists, authorities, and whizzes: Mr. Snowshoe Hare and Mr. Deer Mouse, two indisputable experts—on many different and varied topics, including climate.

The following is a conversation between Mr. Snowshoe Hare and Mr. Deer Mouse

At the sunset hour, the forest is still; overhead, the sky is grey with clouds spitting rain, soaking the ground, and turning it into a muddy, sticky mess. What is now an open field and not that place of the past that normally had been scented with the smack of Douglas fir and Ponderosa pine trees was separated and isolated by what was a natural grassland area. Within what had been the grassy area filled with Oregon grape, snowberry and chokeberry punctuated by shades of lime-green-stemmed yellow, monkey flowers bordered what was the peacock blue river, not at all blue now. A river passed beneath what had been a wooden footbridge but now is a scattered jack-straw-like pile of broken and charred timbers. And on the other side of what used to be the pristine river were the burnt remains of the outhouses and three-sided shelters that had been the cozy campsite refuges on the opposite shore of the river. The remote campsites had been sites to relax in, to be one with Nature, to rest after an arduous day of hiking. The entire place was—had been—accompanied by a steady release of pent-up stress accumulated by being associated with today's human race. Or, in this case, an association with the animal world . . . those by choice, of course. However, even before the climatic catastrophe altered this landscape into a battlefield-like scene, a close association with Mr. Grizzly Bear, Mr. Wolverine, Mr. Cougar, and a few other Mr.'s and/or Ms.'s is not and was not recommended.

Yes, this is what this place is today. With its cold malevolent air hovering above the naked, muddy ground before its transformation to what could be described as a battlefield scene of death and destruction—no movement, no wind, no people. It's like they just vanished or were never there to begin with—that is, except for two stragglers.

"Mr. Deer Mouse, so good to see you again," says Mr. Snowshoe Hare while craning his neck to the left, to the right, and back and forth repeatedly—this is a constant maneuver for Mr. Hare, because unlike the rabbit, hares are not able to view front and back and all around with an almost perfect 360-degree range of sight. Being alert for hungry foxes, wolves, coyotes, bobcats, and the deadly wolverine is just a normal and continuous and lifesaving and necessary habit. As far as Mr. Snowshoe Hare is concerned, Mr. Fox, Mr. Wolf, Mr. Coyote, and Mr. Bobcat can look elsewhere for their culinary delight. (And truth be told, that is exactly what all those critters, including human critters, were doing; they were/are looking elsewhere.) With the cover of darkness in the forest (that is, what was once a forest) come the nocturnal predators, who seem more in need of nourishment than rest at night. However, Mr. Hare has noticed recently that it is difficult to spot any of those who would love to eat him—those types seemed to have moved on to somewhere else.

They have moved on.

Mr. Mouse pauses a bit before saying, "Well, Mr. Hare, it is good to see you, too." Why the pause? Well, Mr. Mouse noticed that Mr. Hare appears to have the same problem he has: a loss of weight, an obvious change in appearance. Meanwhile, Mr. Mouse also is nervously craning his neck here and there and everywhere for the same reason as Mr. Snowshoe Hare: They are on predator watch, even though the predators are long gone. But for Mr. Mouse, it is not just hungry foxes, wolves, coyotes, and bobcats that he must be on guard for but also the northern hawk owl. At this particular time of the evening, the owl is on the prowl and does not howl; moreover, the owl is camouflaged with its plumage that makes it almost invisible to its victims. Truth be told, the northern hawk owl's beak might water at just the sight of Mr. Mouse and any of his close relatives. Once spotted by the owl, it is time to break out the salt, pepper, and ketchup, at least in the owl's distinctive-looking eyes. Even though it has always been the case that he and his kind must be on constant watch for those who would consume them with or without salt, pepper, or ketchup, there just doesn't seem to be anyone around—Mr. Mouse and Mr. Rabbit do not know yet that they are a couple of their kind who have not gotten the message. Still, one never lets his guard down—no, never. Like humans, animals are creatures of habit. Routine and habit can sometimes keep you safe.

So what is the message that Mr. Mouse and Mr. Rabbit have not gotten yet?

Well, nice of you to ask.

Truth be told, the climate in their indigenous area has changed and is changing more each day, so Mr. Mouse and Mr. Rabbit are only a couple of leftovers who have not moved on to those so-called "greener pastures."

Anyway, still alert as always, Mr. Snowshoe Hare says, "Have you noticed all the changes south of here . . . you know, near what them humans call a city . . . lately . . . actually, recently?"

Mr. Mouse, using one of his white feet to scratch his side, says, "Yes, terrible. . . . I wondered at first what was going on around here. . . . I know all of us animals are prey for the other animals, but it is difficult if not impossible to sight any of our enemies or our old friends . . . they seem to have vanished."

Mr. Hare nods in agreement as he looks here, there, and everywhere, still on guard even though enemy and friend had departed weeks, days, and even hours ago seeking those green pastures to the north. And before Mr. Hare responds to Mr. Mouse's comments, he scans around in the dimming light at the shadows and the scorched earth all around them. Then he looks at his friend and says, "Hard to believe what has happened to this place . . . it is wasted . . . pure waste . . . all gone . . . nothing to eat . . . nothing to drink except for the raindrops and muddy water . . . the smell of smoke is almost choking . . . makes you wonder what happened? And why did it happen, and where is everyone?"

Mr. Mouse listens to Mr. Hare while at the same time trying to look through the darkness at the wasteland surrounding them. Not much to see, he thinks. Also, he feels afraid, scared not by predators lurking but by the landscape itself. Maybe it's the silence. No wind blowing in the trees and boughs touching each other. Not a standing tree anywhere. Maybe it is that now there is too much. When the landscape is wide open without trees or foliage, it seems so big, so scary, so terrifying, so daunting, and so forbidding. Maybe it's the lack of something to cling to, to seek cover under or within that's making him feel lost, disoriented. It scares him because he wonders what will be next. Anyway, shuddering a bit, but brushing himself off, he comments on Mr. Hare's words by saying: "You know, all that is true . . . we both are hungry and have lost weight, and if it were not for the rain we would dry up, shrivel up, shrink even more. We need to find them greener pastures, so to speak, up north where all our friends and others have gone . . . we need to migrate . . . don't you think?"

"What do I think? Well as those humans say, hell's bells and time to ship out . . . for sure. Anyway, you and I seem to be the last of those lingering here in this new wasteland . . . that last forest fire, earthquake, and terrible storm have wiped this place dirty . . . it is disgusting and disturbing, for sure. We will have to move on; right now we are between a burned-down forest and the promised land up north. And when we leave here and go there, we will have two worlds to enter . . . before all this destruction, we had something to eat and to drink and places to hide . . . now nothing. But we will also be entering harm's way . . . making ourselves available to predators . . . as it was before this mess occurred. But if we are to have a chance to survive, we must move on . . . there simply is too little here for us and too many there also . . . that might be the crux of the problem; that is, there are too many for too little wherever we end up. Don't you agree?"

Without pausing, Mr. Mouse replies, "Right on, Mr. Rabbit . . . right on."

1 Setting the Stage

INTRODUCTION

As stated in the prologue, this is not a political book.

This book is about climate and human migration.

Okay, so what does that mean?

It means that I first need to take some time to define climate change migration. I have found from research that the best way to do this is to use an adaptation of Myers (1995) in his explanation of what he calls environmental migration. Consider the following adaptation:

What are climate change refugees?

Climate change refugees are persons who can no longer gain a secure livelihood in their traditional homelands because of what are primarily climate facts (a.k.a. environmental factors) of unusual scope. These factors include drought, deforestation, desertification, soil erosion, and other forms of land degradation; resource deficits such as water resources; decline of urban habitats though massive overloading of city systems (e.g., homeless people setting up camp in the middle of town and disposing of their drug paraphernalia, human waste and garbage on the main streets and sidewalks); emergent problems such as climate change, especially global warming; and natural disasters such as cyclones, storm surges and floods, and earthquakes, with impacts aggravated by human mismanagement. There can be additional factors that exacerbate environmental problems and that derive in part from environmental problems: population growth, widespread poverty, famine, and pandemic disease. In certain circumstances, a number of factors can serve as intermediate drivers of migration, such as major individual accidents and construction disasters (e.g., a dam failure). Note that many of these manifold factors can operate in combination. In the face of climate change problems, people feel they have no alternative but to seek sustenance elsewhere, either within their countries or in other countries and on a semipermanent or permanent basis. Simply stated, the climate change refugee is a person who is displaced within his or her country by certain drivers. There are a number of drivers, but in this book, we avoid drivers that include persecution for reasons of race, religion, sexual orientation, or political opinions and beliefs. Instead, the focus is on environmental drivers such as:

- landlessness
- deforestation
- desertification
- soil erosion
- saline water and water logging of irrigated lands
- water deficiency and droughts
- agricultural stress
- biodiversity depletion

DOI: 10.1201/9781003295211-1

1

- extreme weather events
- population pressure
- disease and malnutrition
- poverty

Climate influences migration. For many, the influence, the connection, the nexus between environmental change (climate change) and human migration, the push-pull effect, and the driver is direct and quite obvious, such as when people experience ruined homes, farms, towns, cities, regions, states, or countries in the wake of drought, flooding, hurricane, tornado, disease (e.g., COVID-19 and its variants), forest fire, sea level rise, relative sea level rise (i.e., rising sea and land subsidence), earthquake, pestilence (crop damage and failures), and other natural or human-driven environmental catastrophes. Sometimes climate's influence is more indirect, such as when a string of forest fires prods a resident to decide it is time to get out of Dodge. Truth be told, the potential link between climate change and migration has been (or is being) widely discussed by policy makers as a driver in the climate-migration nexus. The overall objective of this book is to explore the potential pathways linking climate change and increased migration. This may seem to be an easy link to explain and to describe, but keep in mind that despite the growing concern and focus on climate change and migration, uncertainty remains regarding the pathway linking climate change to migration. In part, this uncertainty is brought about by the inherent complexity of projecting climate change. Climate events can occur spontaneously, so to speak, and migration based on one of these events may not be so obvious and not exactly predictable.

Not exactly predictable? Well, that is true, of course. But also, there is another issue, that of attempting to project population growth and movements that go with the flow. Experts in the field of population growth and the climate change-migration nexus are busy studying climate extremes and migration around the world. Some of these studies are summarized in Table 1.1.

Okay, are the events listed in Table 1.1 naturally occurring? Or are they precipitated by human activity? Isn't this the climate change question of the day, of the week, of the year (2022)? Well, the most recent Intergovernmental Panel on Climate Change (IPCC) Assessment Report states that it is extremely likely that anthropogenic (human-caused) factors are the cause of more than half of the observed increase in the average global surface temperature over the last 60 years (IPCC, 2022). Moreover, IPCC (2022) points out that we have already observed changes in extreme weather events, and it is likely that we can expect to see increases in the length and severity of heat waves and extreme storm events in the future. Note that livelihoods will be changed with the changing patterns of seasonal rainfall. This will also lead to shifts in regional suitability for agriculture (among other things). Keep in mind that projections of future climate change depend on assumptions of greenhouse gases (mainly), and these assumptions are just that: assumptions. While change is inevitable, unavoidable, and almost certain, the various climate models do not agree closely on where and when such changes will occur—climate change is no $2 + 2 = 4$ proposition. There are too many variables, many of which we do not know,

TABLE 1.1
Examples of Climate Extremes Leading to Migration since 2000

Type of Climate Event	Time	Region/Country	Major Impacts of Migration
Drought/soil degradation	2004/2007	Kenya	Increase in temporary labor migration with decreasing soil quality
Heat stress	1991–2013	Pakistan	Increase in long-term migration of men
Forest fires	2010	United States	Increased intention to migrate
Flooding/cyclone	2009	Bangladesh	Increase in male rural-urban migration
Flooding	2011–2012	Pakistan	Increase in rural-urban migration
Drought	2006–2014	Syria	Increase in rural-urban migration
Drought/water scarcity	2005–present	Western Sahel	Increase in labor-related migration of pastoralists
Droughts	1996–present	Peru and Bolivia	Increase in labor-related migration of pastoralists

Sources: Reuveny, 2007; Gray, 2011; Mueller et al., 2014; Nawrotzki et al., 2014; Mallick and Cyclone, 2012; Bhattacharyya and Werz, 2016; Nyong, 2016; Hoffman and Grigera, 2016.

understand, or even notice—this is one of those things we do not know what we do not know about climate change.

Okay, recall the following:

> Some of the scientists, I believe, haven't they been changing their opinion a little bit on global warming? There's a lot of differing opinions and before we react, I think it's best to have a full accounting, full understanding of what's taking place.

> **George W. Bush**
> *Presidential Debate, October 11, 2000*

NOTE TO THE READER

The reader and/or user of this text cannot be expected to work even at the margin of the content presented unless the factors affecting our climate regarding atmospheric pollution are presented first (factors affecting other parts of the environment are discussed later in the text). The focus is on what is affecting today's and tomorrow's climate and its connection to human migration. The basics presented set the foundation for the material that follows. Keep in mind that to attempt to understand present changes to our climate, we must understand past occurrences. Also, we must be familiar with the atmospheric changes observed to date and what caused them.

THE HISTORICAL PAST

Before we begin our discussion of the historical past related to climatic conditions, we need to define the era we refer to when we say "the past." Tables 1.2 and 1.3 are provided to assist us in making this definition. Table 1.2 gives the entire expanse

TABLE 1.2
Geologic Eras and Periods

Era	Period	Millions of Years before Present
Cenozoic	Quaternary	2.5–present
	Tertiary	65–2.5
Mesozoic	Cretaceous	135–65
	Jurassic	190–135
	Triassic	225–190
Paleozoic	Permian	280–225
	Pennsylvanian	320–280
	Mississippian	345–320
	Devonian	400–345
	Silurian	440–400
	Ordovician	500–440
	Cambrian	570–500
Precambrian		4,600–570

TABLE 1.3
Epochs

Epoch	Million Years Ago
Holocene	01–0
Pleistocene	1.6–.01
Pliocene	5–1.6
Miocene	24–5
Oligocene	35–24
Eocene	58–35
Paleocene	65–58

of time from Earth's beginning to the present. Table 1.3 provides the sequence of geological epochs over the past 65 million years, as dated by modern methods. The Paleocene through Pliocene together make up the Tertiary Period; the Pleistocene and the Holocene compose the Quaternary Period.

When we think about climatic conditions in the prehistoric past, two things come to mind—Ice Ages and dinosaurs. Of course, in the immense span of time that prehistory covers, those two eras represent only a moment in time. So let's look at what we know or what we think we know about the past and about earth's climate and conditions. One thing to consider—geological history shows us that the normal climate of the Earth was so warm that subtropical weather reached to 60°N and S, and polar ice was entirely absent.

Only during less than about 1 percent of the Earth's history did glaciers advance and reach as far south as what is now the temperate zone of the northern hemisphere. The latest such advance, which began about 1,000,000 years ago, was marked by geological upheaval and (perhaps) the advent of human life on earth. During this time, vast ice sheets advanced and retreated, grinding their way over the continents.

A TIME OF ICE

Two billion years ago, the oldest known glacial epoch occurred. A series of deposits of glacial origin in southern Canada, extending east to west about 1,000 miles (about the distance from Florida to New York City), show us that within the last billion years or so, apparently at least six major phases of massive, significant climatic cooling and consequent glaciation occurred at intervals of about 150 million years. Each lasted for as long as 50 million years.

Examination of land and oceanic sediment core samples clearly indicates that in more recent times, many alternating episodes of warmer and colder conditions occurred over the last 2 million years (during the middle and early Pleistocene epochs). In the last million years, at least eight such cycles have occurred, with the warm part of the cycle lasting for a relatively short interval.

During the Great Ice Age (the Pleistocene epoch), ice advances began, a series of them that at times covered over one quarter of the earth's land surface. Great sheets of ice thousands of feet thick, these glaciers moved across North America over and over, reaching as far south as the Great Lakes. An ice sheet thousands of feet thick spread over Northern Europe, sculpting the land and leaving behind lakes, swamps, and terminal moraines as far south as Switzerland. Each succeeding glacial advance was more severe than the previous one. Evidence indicates that the most severe began about 50,000 years ago and ended about 10,000 years ago. Several interglacial stages separated the glacial advances, melting the ice. Average temperatures were higher than ours today.

Wait a minute! Temperatures were higher than today? Yes, they were. Think about that as we proceed.

Because one tenth of the globe's surface is still covered by glacial ice, scientists consider the Earth still to be in a glacial stage. The ice sheet has been retreating since the climax of the last glacial advance, and world climates, although fluctuating, are slowly warming.

From our observations and from well-kept records, we know that the ice sheet is in a retreating stage. The records clearly show that a marked worldwide retreat of ice has occurred over the last 100 years. World famous for its fifty glaciers and 200 lakes, Glacier National Park in Montana does not present the same visual experiences it did 100 years ago. In 1937, a 10-foot pole was put into place at the terminal edge of one of the main glaciers. The sign is still in place, but the terminal end of the glacier has retreated several hundred feet back up the slope of the mountain. Swiss resorts built during the early 1900s to offer scenic glacial views now have no ice in sight. Theoretically, if glacial retreat continues, melting all the world's ice supply, sea levels would rise more than 200 feet, flooding many of the world's major cities. New York and Boston would become aquariums.

The question of what causes ice ages is one scientists still grapple with. Theories range from changing ocean currents to sunspot cycles. On one fact we are certain, however; an ice age event occurs because of a change in Earth's climate. But what could cause such a drastic change?

Climate results from uneven heat distribution over earth's surface. It is caused by the Earth's tilt—the angle between the earth's orbital plane around the sun and its rotational axis. This angle is currently 23.5 degrees, but it has not always been that. The angle, of course, affects the amount of solar energy that reaches the earth and where it falls. The heat balance of the Earth, which is driven mostly by the concentration of carbon dioxide (CO_2) in the atmosphere, also affects long-term climate. If the pattern of solar radiation changes, or if the amount of CO_2 changes, climate change can result. Abundant evidence that the earth does undergo climatic change exists, and we know that climatic change can be a limiting factor for the evolution of many species.

Evidence (primarily from soil core samples and topographical formations) tells us that change in climate includes events such as periodic ice ages characterized by glacial and interglacial periods. Long glacial periods lasted up to 100,000 years; temperatures decreased about 9°F, and ice covered most of the planet. Short periods lasted up to 12,000 years, with temperatures decreasing by 5°F and ice covering 40° north latitude and above. Smaller periods (e.g., the "Little Ice Age," which occurred from about 1000–1850 AD) had about a 3°F drop in temperature. *Note*: Despite its name, the Little Ice Age was a time of severe winters and violent storms, not a true glacial period. These ages may or may not be significant, but consider that we are presently in an interglacial period and that we may be reaching its apogee. What does that mean? No one knows with any certainty.

Let's look at the effects of ice ages (i.e., effects we think we know about). Changes in sea levels could occur. Sea level could drop by about 100 meters (about the height of the Statue of Liberty) in a full-blown ice age, exposing the continental shelves. Increased deposition during melt would change the composition of the exposed continental shelves. Less evaporation would change the hydrological cycle. Significant landscape changes could occur—on the scale of the Great Lakes formation. Drainage patterns throughout most of the world and topsoil characteristics would change. Flooding on a massive scale could occur. How these changes would affect you depends on whether you live in Northern Europe, Canada, Seattle, Washington, around the Great Lakes, or near a seashore.

We are not sure what causes ice ages, but we have some theories (don't people always have theories?). To generate a full-blown ice age (massive ice sheet covering most of the globe), scientists point out that certain periodic or cyclic events or happenings must occur. Periodic fluctuations would have to affect the solar cycle, for instance; however, we have no definitive proof that this has ever occurred.

Another theory speculates that periods of volcanic activity could generate masses of volcanic dust that would block or filter heat from the sun, thus cooling down the Earth. Some speculate that the carbon dioxide cycle would have to be periodic or cyclic to bring about periods of climate change. There is reference to a so-called Factor 2 reduction, causing a 7°F temperature drop worldwide. Others speculate that another global ice age could be brought about by increased precipitation at the poles, due to changing orientation of continental land masses. Others theorize that a global ice age would result if changes in the mean temperatures of ocean currents decreased. But the question is how? By what mechanism? Are these plausible theories? No one is sure—this is speculation. Some would say it is feel-good speculation; others say it is real, honest speculation. So which one is it? We have no clue; we are not sure.

Speculation aside, what are the most probable causes of ice ages on Earth? According to the *Milankovitch hypothesis*, ice age occurrences are governed by a combination of factors: (1) the Earth's change of altitude in relation to the Sun (the way it tilts in a 41,000-year cycle and at the same time wobbles on its axis in a 22,000-year cycle), making the time of its closest approach to the Sun come at different seasons and (2) the 92,000-year cycle of eccentricity in its orbit around the Sun, changing it from an elliptical to a near-circular orbit, the most severe period of an ice age coinciding with the approach to circularity.

So what does all this mean? We have a lot of speculation about ice ages and their causes and effects. This is the bottom line. We know that ice ages occurred—we know that they caused certain things to occur (e.g., formation of the Great Lakes), and although there is a lot we do not know, we recognize the possibility of recurrent ice ages. Lots of possibilities exist. Right now, no single theory is sound, and doubtless many factors are involved. Keep in mind that the possibility does exist that we are still in the Pleistocene Ice Age. It may reach another maximum in another 60,000-plus years or so.

WARM WINTER

The headlines we see in the paper sound authoritative: "1997 Was the Warmest Year on Record," "Scientists Discover Ozone Hole Is Larger Than Ever," "Record Quantities of Carbon Dioxide Detected in Atmosphere." Or maybe you saw the one that read "January 1998 Was the Third Warmest January on Record." Other reports indicate we are undergoing a warming trend, but conflicting reports abound. This section discusses what we think we know about climate change.

Two environmentally noteworthy events took place late in 1997: El Niño's return and the Kyoto Conference on Global Warming and Climate Change. News reports blamed El Niño for just about anything that had to do with weather conditions throughout the world. Some incidents were indeed El Niño related or generated: the out-of-control fires, droughts, floods, the stretches of dead coral with no sign of fish in the water, and few birds around certain Pacific atolls. The devastating storms that struck the west coasts of South America, Mexico, and California were also probably El Niño related. El Niño's effect on the 1997 hurricane season, one of the mildest on record, is not in question, either.

Does a connection exist between El Niño and global warming or global climate change? On December 7, 1997, the Associated Press reported that while delegates at the global climate conference in Kyoto haggle over greenhouse gases and emission limits, a compelling question has emerged: "Is global warming fueling El Niño?" Nobody knows for sure because we need more information than we have today.

The data we do have, however, suggest that El Niño is getting stronger and more frequent.

Some scientists fear that the increasing frequency and intensity of El Niño's (records show that two of the last century's three worst El Niños came in 1982 and 1997) may be linked to global warming. At the Kyoto Conference, experts said the hotter atmosphere is heating up the world's oceans, setting the stage for more frequent and extreme El Niños. Weather-related phenomena seem to be intensifying throughout the globe. Can we be sure that this is related to global warming yet? No. Without more data, more time, more science (real science), we cannot be sure.

According to the Associated Press coverage of the Kyoto Conference, scientist Richard Fairbanks reported that he found startling evidence of our need for concern. During two months of scientific experiments conducted in autumn 1997 on Christmas Island, the world's largest atoll in the Pacific Ocean, he witnessed a frightening scene. The water surrounding the atoll was 7°F higher than average for the time of year, which upset the balance of the environmental system. According to Fairbanks, 40 percent of the coral was dead, the warmer water had killed off or driven away fish, and the atoll's plentiful bird population was completely gone.

No doubt, El Niños are having an acute impact on the globe; however, we do not know if these events are caused by or intensified by global warming. What do we know about global warming and climate change? *USA Today* (December 1997) discussed the results of a report issued by the Intergovernmental Panel on Climate Change. They interviewed Jerry Mahlman of the National Oceanic and Atmospheric Administration and Princeton University and presented the following information about what most scientists agree on:

- There is a natural greenhouse effect, and scientists know how it works; without it, Earth would freeze.
- The Earth undergoes normal cycles or warming and cooling on grand scales. Ice ages occur every 20,000 to 100,000 years.
- Globally, average temperatures have risen 1°F in the past 100 years, within the range that might occur normally.
- The level of human-made carbon dioxide in the atmosphere has risen 30 percent since the beginning of the Industrial Revolution in the 19th century and is still rising.
- Levels of human-made carbon dioxide will double in the atmosphere over the next 100 years, generating a rise in global average temperatures of about 3.5°F (larger than the natural swings in temperature that have occurred over the past 10,000 years).
- By 2050, temperatures will rise much higher in northern latitudes than the increase in global average temperatures. Substantial amounts of northern sea ice will melt, and snow and rain in the northern hemisphere will increase.
- As the climate warms, the rate of evaporation will rise, further increasing warming. Water vapor also reflects heat back to Earth.

THE HEAVY HAND OF HUMANS[1]

Humans are altering the environment in dramatic fashions, especially over the past 200 years. Human activities can profoundly affect the environment. These activities are not secret or mysterious; in fact, they are obvious—and most of us take part in some of these activities somehow daily. Let's begin with the activities of humans that are changing our atmosphere. To summarize these activities: (1) [human] industrial activities emit a variety of atmospheric pollutants, (2) the [human] practice of burning large quantities of fossil fuel introduces pollutants into the atmosphere, (3) [human] transportation practices emit pollutants into the atmosphere, (4) [human] mismanagement and alteration of land surfaces (deforestation) lead to atmospheric problems, (5) the [human] practice of clearing and burning massive tracts of vegetation produces atmospheric contaminants, and (6) [human] agricultural practices produce chemicals such as methane that impact the atmosphere. These human-made or produced alterations to the earth's atmosphere have delivered profound effects, including, just to name a few, increased acid precipitation, localized smog events, greenhouse gases, ozone depletion, and increased corrosion of materials induced by atmospheric pollutants (Spellman, 2021).

What exactly should we do?

We should understand the human-made mechanisms at work destroying our environment and what we are collectively doing to our environment—and we must be aware that our environment is finite, not inexhaustible or indestructible. Our environment can be destroyed. We must also clearly identify and understand both the causal and the remedial factors involved. Recognizing one salient point is essential: Life on earth and the nature of earth's atmosphere are connected—linked together. The atmosphere drives earth's climate and determines its suitability for life. We must work to preserve the quality of our atmosphere.

Only through a cool-headed, scientifically intellectual, informed mindset will we be able to solve our environmental dilemma. To save our environment (and ourselves), we must develop a vision of an environmentally healthy world—an accomplishable vision. And it is something we can accomplish.

The following sections discuss those issues relevant to environmental pollution of our atmosphere and air quality on Earth, including global warming, acid precipitation, photochemical smog, and stratospheric ozone depletion.

GLOBAL WARMING (A.K.A. CLIMATE CHANGE)

Humanity is conducting an unintended, uncontrolled, globally pervasive experiment whose ultimate consequences could be second only to nuclear war. The Earth's atmosphere is being changed at an unprecedented rate by pollutants resulting from human activities, inefficient and wasteful fossil fuel use, and the effects of rapid population growth in many regions. These changes are already having harmful consequences over many parts of the globe.

—Toronto Conference Statement, June 1988

The preceding quotation clearly states the issue. But what is global warming (better stated as global climate change)? Is it a long-term rise in the average temperature of Earth? This is the case, even though the geological record shows abrupt climate changes occur from time to time. Here's a second question, one many people use to question the validity of the concept of global warming as an environmental hazard. Is global warming occurring? The answer to this accompanying question is of enormous importance to all life on Earth—and is the subject of intense debate throughout the globe. Again, all the debate for the occurrence of global warming can't dispute the historical record that points out that measurements made in central England, Geneva, and Paris from about 1700 until the present indicate a general downward trend in surface temperature (Spellman, 2021). However, recent records point to the opposite—Global temperatures have been on the rise.

Now, before moving on, we need to make a clear distinction between the terms global warming and global climate change. Yes, there is a difference between the two. Global warming refers to the observed warming of the planet due to human-caused emissions of greenhouse gases. Climate change refers to all the various long-term changes in our climate, including extreme weather, sea-level rise, and acidification of our oceans (Spellman, 2021).

DID YOU KNOW?

With regard to biodiversity, climate change can have broad effects on it (the number and variety of plant and animal species in a particular location). Although species have adapted to environmental change for millions of years, a quickly changing climate could require adaption on larger and faster scales than in the past. Those species that cannot adapt are at risk of extinction. Even the loss a single species can have cascading effects because organisms are connected through food webs and other interactions (USEPA, 2010).

For the sake of discussion, let's assume that global warming is occurring. With this assumption in place, we must ask other questions, ones that deal with why, how, and what. (1) Why is global warming occurring? (2) How can we be sure it is occurring? (3) What will be the ultimate effects? (4) What can and are we going to do about it? These questions are difficult to answer. The real danger is that we may not be able to definitively answer these questions before it is too late—when we've reached the point that the process has progressed beyond the power of humans to effect prevention or mitigation. This situation raises a red flag—a huge red flag—and raises additional questions. Are we to stand by and do nothing? Are we to simply ignore the potential impact of this problem? Are we to take the consequences of global warming and global climate change lightly? Are we not to take precautionary actions now instead of later—much later, when it is too late? Indeed, a red flag has been raised (a cause-and-effect relationship to the greenhouse effect), but there is still time before it begins to wave in the climate change that is inevitable—when mitigation becomes harder, more expensive, and impossible to effect.

DID YOU KNOW?

With regard to oceans, they and the atmosphere are constantly interacting—exchanging heat, water, gases, and particles. As the atmosphere warms, the ocean absorbs some of this heat. The amount of heat stored by the ocean affects the temperature of the ocean both at the surface and at great depths. Warming of the Earth's oceans can affect and change the habitat and food supplies for many kinds of marine life—from plankton to polar bears. The oceans also absorb carbon dioxide from the atmosphere. Once it dissolves in the ocean, carbon dioxide reacts with sea water to form carbonic acid. As people put more carbon dioxide into the atmosphere, the oceans absorb some of this extra carbon dioxide, which leads to more carbonic acid. An increasingly acidic ocean can have negative effects on animal life, such as coral reefs.

Exactly what is the nature of the problem of global warming and climate change? Hang on. I may not provide all the answers, but we are about to launch into a discussion of the entire phenomena and its potential impact on Earth (Spellman, 2021).

GREENHOUSE EFFECT

The greenhouse effect makes life on Earth as we know it possible. The basic science of greenhouse effect has been understood for more than a century. To understand Earth's greenhouse effect, here's an explanation most people (especially gardeners) are familiar with. In a garden greenhouse, the glass walls and ceilings are transparent to shortwave radiation from the sun, which is absorbed by the surfaces and objects inside the greenhouse. Once absorbed, the radiation is transformed into longwave (infrared) radiation (heat), which is radiated back from the interior of the greenhouse. But the glass does not allow the longwave radiation to escape, instead absorbing the warm rays. With the heat trapped inside, the interior of the greenhouse becomes much warmer than the air outside.

With the earth and its atmosphere, much the same greenhouse effect takes place. The shortwave and visible radiation that reaches earth is absorbed by the surface as heat. The long heat waves are then radiated back out toward space, but the atmosphere instead absorbs many of them. This is a natural and balanced process, and indeed is essential to life on earth. The problem comes when changes in the atmosphere radically alter the amount of absorption and therefore the amount of heat retained. Scientists, in recent decades, speculate that this may have been happening as various air pollutants have caused the atmosphere to absorb more heat.

That this phenomenon takes place at the local level with air pollution, causing heat islands in and around urban centers, is not questioned. The main contributors to this effect are the greenhouse gases: water vapor, carbon dioxide, carbon monoxide, methane, volatile organic compounds (VOCs), nitrogen oxides, chlorofluorocarbons (CFCs), and surface ozone. These gases delay the escape of infrared radiation from the earth into space, causing a general climatic warming. Note that scientists stress that this is a natural process—indeed, the earth would be 33°C cooler than it is presently if the "normal" greenhouse effect did not exist (Spellman, 2021).

DID YOU KNOW?

With regard to forests, although some of them may derive near-term benefits from an extended growing season, climate change is also expected to encourage wildfires by extending the length of the summer fire season. Larger periods of hot weather could stress trees and make them more susceptible to wildfires, insect damage, and disease. Climate change has likely already increased the size and number of forest fires, insect outbreaks, and tree deaths, particularly in Alaska and the West. The area burned in western U.S. forests from 1987 to 2003 is almost seven times larger than the area burned from 1970 to 1986. In the last 30 years, the length of the wildfire season in the West has increased by 78 days.

The problem with Earth's greenhouse effect is that human activities are now rapidly intensifying this natural phenomenon, which may lead to global warming. There is much debate, confusion, and speculation about this potential consequence. Scientists are not entirely sure, nor do they agree about, whether the recently perceived worldwide warming trend is because of greenhouse gases or from some other cause, or whether it is simply a wider variation in the normal heating and cooling trends they have been studying. But if it continues unchecked, the process may lead to significant global warming, with profound effects. Human impact on the greenhouse effect is real; it has been measured and detected. The rate at which the greenhouse effect is intensifying is now more than five times what it was during the last century (Spellman, 2021).

GREENHOUSE EFFECT AND GLOBAL WARMING

Those who support the theory of global warming base their assumptions on humans altering Earth's normal greenhouse effect, which provides necessary warmth for life, keeping the planet as much as 60°F warmer than it otherwise would be—ideal for humans. They blame human activities (burning of fossil fuels, deforestation, and use of certain aerosols and refrigerants) for the increased quantities of greenhouses gases. These gases have increased the amounts of heat trapped in the Earth's atmosphere, gradually increasing the temperature of the whole globe.

Many scientists note that (based on recent or short-term observation) the last decade has been the warmest since temperature recordings began in the late 19th century. They further note that the more general rise in temperature in the last century has coincided with the Industrial Revolution, with its accompanying increase in the use of fossil fuels. Other evidence supports the global warming theory. For example, in the Arctic and Antarctica, places synonymous with ice and snow, we see evidence of receding ice and snow cover. And significantly, as pointed out by NOAA (2021) on February 26, 2021, JPSS polar satellite watched a city-sized chunk of ice calve off (break off) from the northern section of Antarctica's Brunt ice shelf. The iceberg, named A-74, measures 30 nautical miles (34.5 miles) on its longest axis and 18 nautical miles (20.7) on its widest axis. The total area is roughly 1,270 square kilometers (490 square miles), which is similar in size to the area covered by the city of Los Angeles.

Taking a long-term view, scientists look at temperature variations over thousands or even millions of years. Having done this, they cannot definitively show that global warming is anything more than a short-term variation in Earth's climate. They base this assumption on historical records that show the Earth's temperature does vary widely, growing colder with ice ages and then warming again. On another side of the argument, some people point out that the 1980s saw nine of the twelve warmest temperatures ever recorded, and the Earth's average surface temperature has risen approximately 0.6°C (1°F) in the last century. But at the same time, still others offer as evidence that the same decade also saw three of the coldest years: 1984, 1985, and 1986. So what is really going on? We are not certain. But let's assume that we are indeed seeing long-term global warming. If this is the case, we must determine what is causing it. But here we face a problem. Scientists cannot be sure of the greenhouse effect's causes. Global warming may simply be part of a much longer trend of warming since the last ice age. Though much has been learned in the past two centuries of science, little is known about the causes of worldwide global cooling and warming that have sent the earth through a succession of major ice ages and smaller ones. We simply don't have the enormously long-term data to support our theories.

FACTORS INVOLVED WITH GLOBAL WARMING/COOLING

Right now, scientists can point to five factors that could be involved in long-term global warming and cooling.

1. Long-term global warming and cooling could result if changes in the earth's position relative to the sun occur (the earth's orbit around the sun), with higher temperatures when the two are closer together and lower when further apart.
2. Long-term global warming and cooling could result if major catastrophes occur (meteor impacts or massive volcanic eruptions) that throw pollutants into the atmosphere that can block out solar radiation.
3. Long-term global warming and cooling could result if changes in albedo (reflectivity of earth's surface) occur. If the earth's surface were more reflective, for example, the amount of solar radiation radiated back toward space instead of absorbed would increase, lowering temperatures on earth.
4. Long-term global warming and cooling could result if the amount of radiation emitted by the sun changes.
5. Long-term global warming and cooling could result if the shape and relationship of the land and oceans change.

If the composition of the atmosphere changes—this possibility, of course, relates directly to our present concern: Have human activities had a cumulative impact large enough to affect the total temperature and climate of earth? We are not certain right now. We are concerned and alert to the problem, but we are not certain. So the question is: What are we doing about global warming? We answer this pertinent question in the next section.

How Is Climate Change Measured?

Worldwide, scientists are trying to establish ways to test or measure whether green-house-induced global warming is occurring. Scientists are currently looking for signs that collectively are called a greenhouse "signature" or "footprint." If it is occurring, eventually it will be obvious to everyone—but what we really want is clear advance warning. Thus, scientists are currently attempting to collect and then decipher a mass of scientific evidence to find those signs to give us clear advance warning. According to Franck and Brownstone (1992), these signs are currently believed to include changes in:

- **Global temperature patterns**, with continents being warmer than oceans, lands near the Arctic warming more than the tropics, and the lower atmosphere warming while the higher stratosphere becomes cooler.
- **Atmospheric water vapor**, with increasing amounts of water evaporating into the air because of the warming, more in the tropics than in the higher latitudes. Since water vapor is a "greenhouse gas," this would intensify the warming process.
- **Sea surface temperature**, with a uniform rise in the temperature of oceans at their surface and an increase in the temperature differences among oceans around the globe.
- **Seasonality**, with changes in the relative intensity of the seasons, with the warming effects especially noticeable during the winter and in higher latitudes (p. 143).

In a measured, scientific way, these signs give a general overview of some of the changes that would be expected to occur with global warming. Note, however, that from a viewpoint of life on earth, changes resulting from long-term global warming would be drastic—profoundly serious. The most dramatic—and the effect with the most far-reaching results—would be sea level rise.

Acid Precipitation

In the evening, when you stand on your porch during a light rain and look out on your terraced lawn and that flourishing garden of perennials, you feel a sense of calm and relaxation hard to describe—but not hard to accept. The sound of raindrops against the roof of the house and porch, against the foliage and lawn, the sidewalk, the street, and that light wind through the boughs of the evergreens soothes you. Whatever it is that makes you feel this way, rainfall is a major ingredient.

But someone knowledgeable and/or trained in environmental science might take another view of such a welcome and peaceful event. We might wonder to ourselves whether the rainfall is as clean and pure as it should be. Is this rainfall? Or is it rain carrying acids as strong as lemon juice or vinegar with it capable of harming both living and nonliving things like trees, lakes, and human-made structures? This may seem strange to some folks who might wonder why anyone would be concerned about such an off-the-wall matter.

Such an interest was off the wall before the Industrial Revolution, but today the purity of rainfall is a major concern for many people, especially the levels of acidity. Most rainfall is slightly acidic because of decomposing organic matter, the movement of the sea, and volcanic eruptions, but the principal factor is atmospheric carbon dioxide, which causes carbonic acid to form. *Acid rain* (pH < 5.6) (in the pollution sense) is produced by the conversion of the primary pollutants sulfur dioxide and nitrogen oxides to sulfuric acid and nitric acid, respectively. These processes are complex and depend on the physical dispersion processes and the rates of the chemical conversions.

Contrary to widespread belief, acid rain is not a new phenomenon, nor does it result solely from industrial pollution. Natural processes such as volcanic eruptions and forest fires produce and release acid particles into the air. The burning of forest areas to clear land in Brazil, Africa, and other areas also contributes to acid rain. However, the rise in manufacturing that began with the Industrial Revolution dwarfs all other contributions to the problem (Spellman, 2021).

The main culprits are emissions of sulfur dioxide from the burning of fossil fuels, such as oil and coal, and nitrogen oxide, formed mostly from internal combustion engine emissions, which is readily transformed into nitrogen dioxide. These mix in the atmosphere to form sulfuric acid and nitric acid.

In dealing with atmospheric acid deposition, the Earth's ecosystems are not completely defenseless; they can deal with a certain amount of acid through natural alkaline substances in soil or rocks that *buffer* and neutralize acid. The American Midwest and southern England are areas with highly alkaline soil (limestone and sandstone) that provide some natural neutralization. Areas with thin soil and those laid on granite bedrock, however, have little ability to neutralize acid rain.

Scientists continue to study how living beings are damaged and/or killed by acid rain. This complex subject has many variables. We know from various episodes of acid rain that the pollution can travel over long distances. Lakes in Canada, Maine, and New York feel the effects of coal burning in the Ohio Valley. For this and other reasons, the lakes of the world are where most of the scientific studies have taken place. In lakes, the smaller organisms often die off first, leaving the larger animals to starve to death. Sometimes the larger animals (fish) are killed directly; as lake water becomes more acidic, it dissolves *heavy metals*, leading to concentrations at toxic and often lethal levels. Have you ever wandered up to the local lakeshore and observed thousands of fish belly-up? Not a pleasant sight or smell, is it? Loss of life in lakes also disrupts the system of life on the land and the air around them.

DID YOU KNOW?

While acidification of the U.S. and Canadian lakes mentioned previously is of major concern to environmentalists and others, it should be pointed out that one quarter of the carbon dioxide humans emit into the air gets absorbed in the oceans. The carbon dioxide that dissolves in seawater forms carbonic acid, which in turn acidifies the ocean. The point is the pH of the oceans has steadily dropped since the Industrial Revolution. This is an ongoing trend that is not good, especially for marine life.

In some parts of the United States, the acidity of rainfall has fallen well below 5.6. In the northeastern United States, for example, the average pH of rainfall is 4.6, and rainfalls with a pH of 4.0, which is 100 times more acidic than distilled water, are not unusual.

Despite intensive research into most aspects of acid rain, scientists still have many areas of uncertainty and disagreement. That is why progressive, forward-thinking countries emphasize the importance of further research into acid rain.

PHOTOCHEMICAL SMOG

As mentioned earlier, when various hydrocarbons, oxides of nitrogen, and sunlight come together, they can initiate a complex set of reactions that produce several secondary pollutants known as photochemical oxidants or photochemical smog. Smog production is a localized problem, and with regard to its impact on global climate change and global warming, it is production specific to that area—again, meaning that it is a problem, but generally it is a local issue to deal with. *Photochemical smog* of the type most people are familiar with was first noticed in Los Angeles in the early 1940s. Determining its true cause has taken many years. At first it was thought to arise from dust and smoke emitted from factories and incinerators. Accordingly, Los Angeles County officials issued a ban on all outdoor burning of trash and initiated steps toward control of industrial smoke emission. Before long, though, county authorities determined that their initial efforts were not working; the smog continued unabated. Then they went after another suspected culprit, sulfur dioxide (SO_2) given off by oil refineries and by combustion of sulfur-bearing coal. So they placed controls on sulfur dioxide emissions—but still gained no benefit. However, eventually, the reducing of vehicle emissions in Los Angeles County from 1973 to 2017 has shown some beneficial improvement in lowering daily smog levels in the area (Spellman, 2021).

The biochemist Dr. Arie Haagen-Smit, during research aimed at finding the compounds responsible for the pleasant tastes and odors of fruit, by chance found the cause of the smog problem, which showed beyond a doubt that the internal combustion engine was the principal source.

How do internal combustion engines produce smog? A few of the finer details are yet unclear, but the following, in simplified form, appears to be what happens. Smog begins with the elevated temperatures in the internal combustion engine, which cause atmospheric oxygen and nitrogen to react, producing nitric oxide ($N_2 + O_2 = 2NO$). At the same time, varying quantities of fuel in the engine fail to burn completely. This results in a mixture of aldehydes, ketones, olefins, and aromatic hydrocarbons that are expelled in the exhaust. The exhaust enters the atmosphere, where ultraviolet radiation from the sun causes a complex series of reactions to take place. These reactions involve atmospheric oxygen, nitric oxide, and organic compounds. The resulting nitrogen dioxide (NO_2) and ozone (O_3) are formed, both of which are highly toxic and irritating. In addition, this reaction also causes the formation of other constituents of photochemical smog, including formaldehyde, peroxybenzohl nitrate, peroxyacetyl nitrate (PAN), and acrolein.

Photochemical smog is known to cause many annoying respiratory effects—coughing; shortness of breath; airway constriction; headache; chest tightness; and eye, nose, and throat irritation (Masters, 1991).

STRATOSPHERIC OZONE DEPLETION

Ozone (discussed earlier) is formed in the stratosphere by radiation from the sun and helps to shield life on earth from some of the sun's potentially destructive ultraviolet (UV) radiation.

In the early 1970s, scientists suspected that the ozone layer was being depleted. By the 1980s, it became clear that the ozone shield was indeed thinning in some places and at times even has a seasonal hole in it, notably over Antarctica. The exact causes and extent of the depletion are not yet fully known, but most scientists believe that various chemicals in the air are responsible.

Most scientists identify the family of chlorine-based compounds, most notably *chlorofluorocarbons* and chlorinated solvents (carbon tetrachloride and methyl chloroform), as the primary culprits involved in ozone depletion. Molina and Rowland (1974) hypothesized that CFCs containing chlorine were responsible for ozone depletion. They pointed out that chlorine molecules are highly active and readily and continually break apart the three-atom ozone into the two-atom form of oxygen generally found close to Earth in the lower atmosphere.

The Interdepartmental Committee for Atmospheric Sciences (1975) estimated that a 5 percent reduction in ozone could result in a 10 percent increase in cancer. This already frightening scenario was made even more frightening in1987 when evidence showed that CFCs destroy ozone in the stratosphere above Antarctica every spring. The ozone hole had become larger, with more than half of the total ozone column wiped out and all ozone disappearing from some regions of the stratosphere (Davis and Crowell, 1991; Zurer, 1988; Spellman, 2021).

In 1988, Zurer reported that on a worldwide basis, the ozone layer shrank approximately 2.5 percent in the preceding decade. This obvious thinning of the ozone layer, which causes increased chances of skin cancer and cataracts, is also implicated in suppression of the human immune system and damage to other animals and plants, especially aquatic life and soybean crops. The urgency of the problem spurred the 1987 signing of the *Montreal Protocol* by 24 countries, which required signatory countries to reduce their consumption of CFCs by 20 percent by 1993 and 50 percent by 1998, marking a significant achievement in solving a global environmental problem.

Ozone: The Jekyll and Hyde of Chemicals

In Robert Lewis Stevenson's classic horror novel, *Dr. Jekyll and Mr. Hyde*, Jekyll and Hyde are two aspects of the same person. Dr. Jekyll's kind, compassionate character is countered by Mr. Hyde's evil, dispassionate nature. The chemical ozone has the same potential for good and evil within a single entity.

Ozone (O_3) is a molecule containing three atoms of oxygen. In the earth's stratosphere, about 50,000 to 120,000 feet high, ozone molecules band together to form a protective layer that shields the earth from some of the sun's potentially destructive ultraviolet radiation. Stratospheric ozone (ozone in its kindly Dr. Jekyll incarnation), formed in the atmosphere by radiation from the sun, provides us with an enormously beneficial function. Life as we know it on earth could have evolved only with the protective ozone shield in place.

The Centers for Disease Control in Atlanta looks at ozone more critically, however. They point out that ozone (in its evil Mr. Hyde form) is an extraordinarily dangerous pollutant. Only two-hundredths of a gram of ozone are a lethal dose. A single 14-ounce aerosol can filled with ozone could kill 14,000 people. Ozone is nearly as effective at destroying lung tissue as mustard gas. Not only is ozone a poisonous gas for us on earth, but it is also a main contributor to air pollution, especially smog.

CLIMATE CHANGE MIGRATION[2]

After having presented some of the issues related to atmospheric pollution and its effect on climate change, it is important to point out that climate change and climate variability establish the risk of serious negative impacts on environmental and human structures or systems, including extreme events such as heat waves, droughts, storms, floods, wildfires, changing rainfall, sea-level rise, increased salinization, decreased soil fertility, and others (IPCC, 2015). Damage or loss of land and property due to these events could lead to population displacement as migrants relocate (Dun and Gemenne, 2008; Morton et al., 2008). Again, the specific ways in which climate change impacts migration are not a 2 + 2 = 4 solution or proposition; they bring about disagreement in several areas of discussion and thought. Think about it like this: if you own a farm and some type of climate-driven disaster wipes out your crops not once but several times over the past 2 or 3 years and you decide to get out of Dodge and to migrate to greener pastures (wherever that might be), is the driver the environment, or is it a lack of economic opportunity? This is one of the complexities that make the comparisons with the equation 2 + 2 = 4 doubtful, uncertain, or shaky to a point.

The problem with providing exact data and/or rationale for climate change migration is that it is a long-term, enduring, and long-standing process. Because of the absence of certain provenance of extreme events to climate change, there is a lack of data and instruments to directly document impacts of climate change by itself on migration. In its place, it's all about shorter-term climate variability; that is, most work in this area examines impacts of in-your-face, in-your-own-backyard events such as climate extremes like flooding related to sea-level rise, forest fires, extreme storm events, and so forth. The point is, we do not know what we do not know about long-term climate change on population displacement in the terms of migration. Another point to consider is that if we wait for climate-change–induced impacts to be clear, it may be too late to take preventive action(s) (Costello et al., 2009).

With regard to the number of environmental migrants, estimates of the numbers of people being displaced by climate change (including both short-term and long-term movements) by 2050 range from 50 million to 1 billion people (Adger et al., 2015; Stern, 2007; Human Tide, 2022). Note that Jacobson put forward the low end of the range (50 million environmental refugees) (a.k.a. climate change refugees) as an estimate of populations "at risk" of a 1-meter sea level rise. However, the most-cited number (but not rigorously tested), 200 million, is based on projections conducted by Myers in 1995 and 2000. These projections are consistent with conservative estimates of climate impacts, but there is uncertainty in these estimates, and it is necessary to use extrapolations due to lack of data (Black, 2016; Myers, 1995, 2005; Brown, 2015; Stern, 2007).

THE BOTTOM LINE

For the purposes of this book, I accept the premise, the presumption, and the extrapolations that climate change is a driver for migration—and I fully support the United Nations (2017) view that climate change is a key driver of migration. In thinking about climate change and its repercussions, keep in mind that people in general often react to difficult-to-deal-with information emotionally, not logically, and news of pending environmental crisis is no exception. Those reactions cover a wide range of conditions, from denial to hysteria, to indifference, to obsession, to anger, to activism, to foolhardiness, and none of those reactions—or the people behind them—do much to bring us closer to a solution.

By the time our societal and political systems get through with the information our scientists provide us, public evaluation by anything close to scientific methodology is impossible, because the information the public receives is distorted and incomplete—chopped up and twisted to provide a good sound byte and tipped to suit the political beliefs of those in power and the financial concerns of the owners.

Little doubt exists that we are putting our environment—and thus ourselves—at serious risk. While sometimes we are not certain (yet) of the exact causes—or all the causes—the changes we observe in our world are measurable. As active members of an increasingly global economy, we increase our risk by pretending it will go away. Eventually, we will have to face these problems—and migration may be how we solve them; that is, if there is still anywhere to migrate to.

NOTES

1. Information in the following section is adapted from F. Spellman's (2021) *Climate Change*. Lanham, MD: Bernan Press.
2. The information in this section is based on the National Library of Medicines: *Exploring the Climate Change, Migration and Conflict Nexus* by K. Burrows and P.L. Kinney, accessed 01/25/2022 @ http://pubmed.ncbi.nlm.nih.gov.

REFERENCES

Adger, W.N., et al. (2015). Human security. In Darros, V.R. et al., editors. *Climate Change 2014 Impacts, Adaptation, and Vulnerability. Part A: Global and Sectoral Aspects. Combination of working group II to the Fifth Assessment Report of the Intergovernmental Panel on Climate Change*. Cambridge, UK: Cambridge University Press, pp. 755–791.

Bhattacharyya, A., and Werz, M. (2016). Climate change: Migration, and conflict in South Asia: Rising tensions and policy options across the subcontinent. Accessed 01/26/2022 @ https://www.american progress.org/wp-content/

Black, R. (2016). New issues in refugee research. Accessed 01/27/2022 @ www.unher.ore/ 3ae6a0d00.html.

Brown, O. (2015). Migration and climate change. Accessed 01/27/2022 @ www.iom.ez.files/ migration_and-Climate_Change.

Costello, A., Abbas, M., Allen, A., Ball, S., Bell, S., Bellamy, R., Friel, S., Groce, N., Johnson, A., Kett, M., et al. (2009). Climate change. *Lancet* 373:1693–1733.

Davis, M., and Crowell, L.E. (1991). *Introduction to Environmental Engineering* (2nd ed.). New York: McGraw-Hill.

Dun, O., and Gemenne, F. (2008). Defining "environmental migration". *Forced Migr. Rev.* 31:10–11.

Franck, I., and Brownstone, D.M. (1992). *The Green Encyclopedia*. New York: Wiley.

Gray, C. (2011). Soil quality and human migration in Kenya and Uganda. *Glob. Environ. Chang.* 21:421–430.

Hoffman, M., and Grigera, A. (2016). Conflict in the Amazon and the Andes: Rising tensions and policy options for South America. Accessed 01/26/2022 @ https://americanprogress. org/wp-conrenr/uploads.

Human Tide: The Real Migration Crisis. (2022). Accessed 01/27/2022 @ www.christainaid. org.uk/Images/human-tide.pdf.

Interdepartmental Committee for Atmospheric Sciences. (1975). *The Possible Impact of Fluorocarbons and Halocarbons on Ozone*, ICAS Report no. 18a, Washington, DC, 75 pp.

IPCC Fourth Assessment Report: Synthesis Report. (2015). Accessed 01/26/2022 @ www. ipcc.ch/report/ar4/syr/.

IPCC Fifth Assessment Report: Synthesis Report. (2022). Accessed 01/26/2022 @ www.ipcc. ch/report/ar5/syr/.

Mallick, B., and Cyclone, V.J. (2012). Coastal society and migration: Empirical evidence from Bangladesh. *Int. Dev. Rev.* 34:217–240.

Masters, G.M. (1991). *Environmental Engineering and Science*. Englewood Cliffs, NJ: Prentice Hall.

Molina, M., and Rowland, F. (1974). Stratospheric risks for chlorofluoromethane. *Nature* 249:810–812.

Morton, A., Boncour, P., and Lackso, F. (2008). Human security policy challenges. *Forced Migr. Rev.* 31:5–8.

Mueller, V., Gray, C., and Kosec, K. (2014). Heat stress increases long-term human migration in rural Pakistan. *Nat. Clim. Chang.* 4:182–185.

Myers, N. (1995). *Environmental Exodus: An Emergent Crisis in the Global Arena*. Washington, DC: Climate Institute.

Myers, N. (2005). *Environmental Refugees: An Emergent Security Issue; Documents from the 13th Economic Forum*. Proceedings of the 13th Meeting of the Organization for Security and Co-operation in Europe (OSCE) Economic Forum, Section III, Prague, Czech Republic, 23–27 May.

Nawrotzki, R.J., Brenkert-Smith, H., and Hunter, L. (2014). Wildfire-migration dynamics: Lessons from Colorado's four-mile canyon fire. *Soc. Nat. Resour.* 27:215–235.

NOAA (National Oceanic and Atmospheric Administration). (2021). Part of Brunt ice shelf Antarctica broke off. Accessed 01/29/2022 @ noaa@nesdis.noaa.gov/large-icebreak-breaks-antarctica.

Nyong, A. (2016). Climate-related conflicts in West Africa. Accessed 01/26/2022 @ https://www.wilsoncenter.org/publication/climate-related-conflicts-west-africa.

Reuveny, R. (2007). Climate change-induced migration and violent conflict. *Political Geogr.* 26:656–773.

Spellman, F.R. (2021). *Climate Change*. Lanham, MD: Bernan Press.

Stern, N. (2007). *The Economics of Climate Change: The Stern Review*. Cambridge: Cambridge University Press.

United Nations. (2017). *United Nations Climate Change Annual Report, 2017.* New York: United Nations.

USEPA. (2010). *Climate Change Indicators*. Washington, DC: U.S. Environmental Protection Agency.

Zurer, P. (1988). Ozone, skin cancer and the SST. Accessed 12/12/21 @ https//www.nas.nasa.gov/about/ozone.

2 Water's Impact on Migration

THE WELL IS DRY[1]

Have you ever given any thought to water?

Have you ever been thirsty?

Have you ever drunk water that made you sick, made you vomit, made you hospitalized or worse?

Do you have even the remotest idea about what water is? What water does? What water is capable of? Where water has been? Where is water going? Where has the water gone?

Well, if you have no way of answering any of these questions, let's get down to the 411 about water.

Whether we characterize it as ice, rainbow, steam, frost, dew, soft summer rain; as fog; as flood or avalanche; or as stimulating as a stream or cascade, water is special—water is strange—water is different.

Water is the most abundant inorganic liquid in the world; moreover, it occurs naturally anywhere on earth. Awash with it, life depends on it, and yet water is also quite different.

Water is scientifically different. With its rare and distinctive property of being denser as a liquid than as a solid, it is different. Water is different in that it is the only chemical compound found naturally in solid, liquid, and gaseous states. Water is sometimes called the *universal solvent*. This is a fitting name, especially when you consider that water is a powerful reagent, which is capable in time of dissolving everything on earth.

Water is different. It is usually associated with all the good things on earth. For example, water is associated with quenching thirst, with putting out fires, and with irrigating the earth. The question is: Can we really say emphatically, definitively, that water is associated with only those things that are good?

Not really!

Remember, water is different; nothing, absolutely nothing, is safe from it.

Water is different. This unique substance is odorless, colorless, and tasteless. Water covers 71 percent of the earth completely. Even the driest dust ball contains 10–15 percent water.

Water and life—life and water—inseparable.

The prosaic becomes wondrous as we perceive the marvels of water.

Three hundred twenty-six million cubic miles of water cover earth, but only 3 percent of this total is fresh, with most locked up in polar ice caps, glaciers, in lakes; in flows through soil; and in river and stream systems back to an increasingly salty sea (only 0.027 percent is available for human consumption). Water is different.

DOI: 10.1201/9781003295211-2

Standing at a dripping tap, water is so palpably wet that one can hear the drip-drop-plop.

Water is special—water is strange—water is different—more importantly, water is critical to our survival, yet we abuse it, discard it, foul it, curse it, dam it, and ignore it. At least this is the way we view the importance of water at this moment in time . . . however, because water is special, strange, and different, the dawn of tomorrow is pushing for quite a different view.

Along with being special, strange, and different, water is also a contradiction, a riddle.

How?

Consider the Chinese proverb that states, "Water can both float and sink a boat."

Saltwater is different from fresh water. Moreover, this text deals with freshwater, the lack of freshwater, the lack of quality water, and the lack of safe drinking water and ignores salt water because saltwater fails its most vital duty, which is to be pure and sweet and serve to nourish us.

The presence of water everywhere feeds these contradictions. Lewis (1996, p. 90) points out that "water is the key ingredient of mother's milk and snake venom, honey and tears."

DID YOU KNOW?

There's a lot of salty water on our planet. By some estimates, if the salt in the ocean could be removed and spread evenly over the Earth's land surface, it would form a layer more than 500 feet (166 meters) thick, about the height of a forty-story office building. The question is: Where did all this salt come from? Stories in folklore and mythology from almost every culture have a story explaining how the oceans became salty. The answer is really quite simple. Salt in the ocean comes from rocks on land. Here's how it works: The rain that falls on the land contains some dissolved carbon dioxide from the surrounding air. This causes the rainwater to be slightly acidic due to carbonic acid. The rain physically erodes the rock, and the acids chemically break down the rocks and carry salts and minerals along in a dissolved state as ions. The ions in the runoff are carried to the streams and rivers and then to the ocean. Many of the dissolved ions are used by organisms in the ocean and are removed from the water. Others are not used up and are left for long periods of time where their concentrations increase over time. The two ions that are present most often in seawater are chloride and sodium. These two make up over 90 percent of all dissolved ions in seawater (USGS, 2013).

Leonardo da Vinci gave us insight into more of water's apparent contradictions:

Water is sometimes sharp and sometimes strong, sometimes acid and sometimes bitter.

Water is sometimes sweet and sometimes thick or thin.

Water sometimes brings hurt or pestilence, sometimes health-giving, sometimes poisonous.

Water suffers changes into as many natures, as are the various places through which it passes.

As with the mirror that changes with the color of its object, so does water alter
with the nature of the place, becoming noisome, laxative, astringent, sulfu-
rous, salty, incarnadined, mournful, raging, angry, red, yellow, green, black,
blue, greasy, fat, or slim.

Water sometimes starts a conflagration; sometimes it extinguishes one.

Water is warm and is cold.

Water carries away or sets down.

Water hollows out or builds up.

Water tears down or establishes.

Water empties or fills.

Water raises itself or burrows down.

Water spreads or is still.

Water is the cause at times of life or death or increase of privation, nourishes at
times and at others does the contrary.

Water at times has a tang; at times it is without savor.

Water sometimes submerges the valleys with a great flood.

In time and with water, everything changes.

Water's contradictions can be summed up by simply stating that though the globe is
awash in it, water is no single thing but an elemental force that shapes our existence.
Leonardo's last contradiction, "In time and with water, everything changes" concerns us most in this text.

Many of Leonardo's water contradictions are apparent to most observers. But with
water there are other factors that do not necessarily stand out, that are not always so
apparent. This is made clear by the following example—what you see on the surface
is not necessarily what lies beneath.

STILL WATER

Consider a river pool, isolated by fluvial processes and time from the mainstream flow.
We are immediately struck by one overwhelming impression: It appears so still . . . so
very still . . . still enough to soothe us. The river pool provides a poetic solemnity, if
only at the pool's surface. No words of peace, no description of silence or motionless-
ness can convey the perfection of this place, in this moment stolen out of time.

We ask ourselves, "The water is still, but does the term 'still' correctly describe
what we are viewing . . . is there any other term we can use besides still—is there
any other kind of still?"

Yes, of course, we know many ways to characterize still. For sound or noise, "still" can
mean inaudible, noiseless, quiet, or silent. With movement (or lack of movement), still can
mean immobile, inert, motionless, or stationary. At least, this is how the pool appears to
the casual visitor on the surface. The visitor sees no more than water and rocks.

How is the rest of the pool? We know very well that a river pool is more than just a
surface. How does the rest of the pool (for example, the subsurface) fit the descriptors
we tried to use to characterize its surface? They fit; they do not. In time, we will go
beneath the surface, through the liquid mass, to the very bottom of the pool to find
out. For now, remember that images retained from first glances are always incorrectly perceived, incorrectly discerned, and never fully understood.

On second look, we see that the fundamental characterization of this pool's surface is correct enough. Wedged in a lonely riparian corridor—formed by riverbank on one side and sand bar on the other—between a youthful, vigorous river system on its lower end and a glacier- and artesian-fed lake on its headwater end, entirely overhung by mossy old Sitka spruce, the surface of the large pool, at least at this location, is indeed still. In the proverbial sense, the pool's surface is as still and as flat as a flawless sheet of glass.

The glass image is a good one, because, like perfect glass, the pool's surface is clear, crystalline, unclouded, transparent, yet perceptively deceptive as well. The water's clarity, accentuated by its bone-chilling coldness, is apparent at close range. Further back, we see only the world reflected in the water—the depths are hidden and unknown. Quiet and reflective, the polished surface of the water perfectly reflects in mirror-image reversal the spring greens of the forest at the pond's edge, without the slightest ripple. Up close, looking straight into the bowels of the pool, we are struck by the water's transparency. In the motionless depths, we do not see a deep, slow-moving reach with muddy bottom typical of a river or stream pool; instead, we clearly see the warm variegated tapestry of blues, greens, blacks stitched together with threads of fine, warm-colored sand that carpets the bottom, at least 12 feet below. Still waters can run deep.

No sounds emanate from the pool. The motionless, silent water doesn't, as we might expect, lap against its bank or bubble or gurgle over the gravel at its edge. Here, the river pool, held in temporary bondage, is patient, quiet, waiting, withholding all signs of life from its surface visitor.

Then the reality check: the present stillness, like all feelings of calm and serenity, could be fleeting, momentary, temporary, you think. And you would be correct, of course, because there is nothing still about a healthy river pool.

At this exact moment, true clarity is present, it just needs to be perceived . . . and it will be.

We toss a small stone into the river pool and watch the concentric circles ripple outward as the stone drops through the clear depths to the pool bottom. For a brief instant, we are struck by the obvious: the stone sinks to the bottom, following the laws of gravity, just as the river flows according to those same inexorable laws— downhill in its search for the sea. As we watch, the ripples die away, leaving as little mark as the usual human lifespan creates in the waters of the world, then disappears as if it had never been. Now the river water is as before, still. At the pool's edge, we look down through the massy depth to the very bottom—the substrate.

We determine that the pool bottom is not flat or smooth but instead is pitted and mounded occasionally with discontinuities. Gravel mounds alongside small corresponding indentations—small, shallow pits—make it apparent to us that gravel was removed from the indentations and piled into slightly higher mounds. From our topside position, as we look down through the cool, quiescent liquid, the exact height of the mounds and the depth of the indentations is difficult for us to judge; our vision is distorted through several feet of water.

However, we can detect near the low gravel mounds (where female salmon buried their eggs, and where their young grow until they are old enough to fend for themselves), and through the gravel mounds, movement—water flow—an upwelling of groundwater. This water movement explains our ability to see the variegated color of pebbles. The mud and silt that would normally cover these pebbles have been washed away by the water's subtle, inescapable movement. Obviously, in the depths, our still water is not as still as it first appeared.

The slow, steady, inexorable flow of water in and out of the pool, along with the up-flowing of groundwater through the pool's substrate and through the salmon redds (nests) is only a small part of the activities occurring within the pool, including the air above it, the vegetation surrounding it, and the damp bank and sandbar forming its sides.

Let's get back to the pool itself. If we could look at a cross-sectional slice of the pool, at the water column, the surface of the pool may carry those animals that can walk on water. The body of the pool may carry rotifers, protozoa, and bacteria—tiny, microscopic animals—as well as many fish. Fish will also inhabit hidden areas beneath large rocks and ledges to escape predators. Going down further in the water column, we come to the pool bed. This is called the benthic zone, and certainly the greatest number of creatures live here, including larvae and nymphs of all sorts, worms, leeches, flatworms, clams, crayfish, dace, brook lampreys, sculpins, suckers, and water mites.

We need to go down even further, down into the pool bed, to see the whole story. How far this goes and what lives here, beneath the water, depends on whether it is a gravelly bed or a silty or muddy one. Gravel will allow water, with its oxygen and food, to reach organisms that live underneath the pool. Many of the organisms that are found in the benthic zone may also be found underneath, in the hyporheal zone.

But to see the rest of the story, we need to look at the pool's outlet and where its flow enters the main river. This is the riffles—shallow places where water runs fast and is disturbed by rocks. Only organisms that cling very well, such as net-winged midges, caddisflies, stoneflies, some mayflies, dace, and sculpins can spend much time here, and the plant life is restricted to diatoms and small algae. Riffles are a good place for mayflies, stoneflies, and caddisflies to live because they offer plenty of gravel to hide.

At first, we struggled to find the "proper" words to describe the river pool. Eventually, we settled on "still waters." We did this because of our initial impression and because of our lack of understanding—lack of knowledge. Even knowing what we know now, we might still describe the river pool as still waters. However, we must call the pool what it really is: a dynamic habitat. This is true, of course, because each river pool has its own biological community, all members interwoven with each other in complex fashion, all depending on each other. Thus, our river pool habitat is part of a complex, dynamic ecosystem. On reflection, we realize, moreover, that anything dynamic certainly can't be accurately characterized as "still"—including our river pool.

You have not had the opportunity to observe a river pool like the one described here. Such an opportunity does not interest you. However, the author's point can be made in a different manner.

Take a moment out of your hectic schedule and perform an action most people never think about doing. Hold a glass of water (like the one in Figure 2.1) and think about the substance within the glass—about the substance you are getting ready to

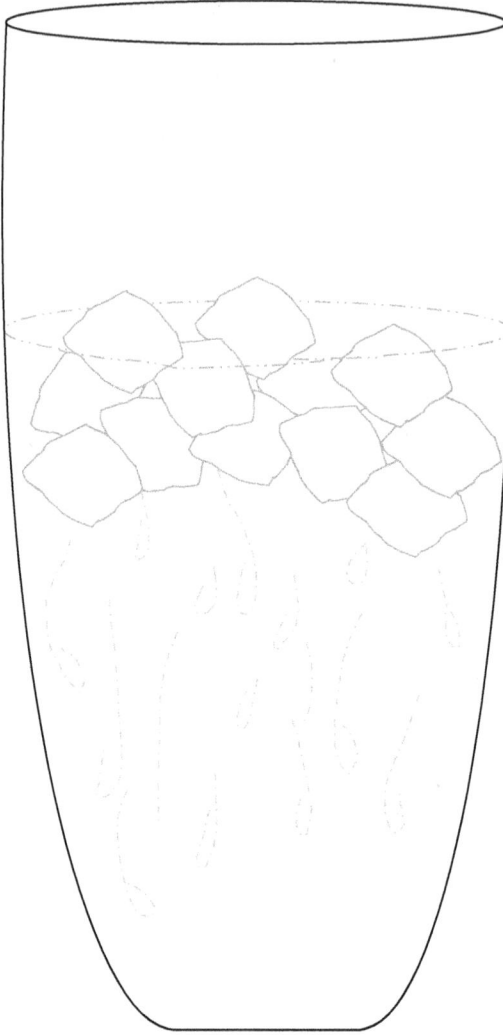

FIGURE 2.1 A glass of iced drinking water.

drink. You are aware that the water inside a drinking glass is not one of those items people usually spend much thought on unless they are tasked with providing the drinking water—or dying of thirst.

As mentioned earlier, water is special, strange, and different. Some of us find water fascinating—a subject worthy of endless interest because of its unique behavior, limitless utility, and ultimate and intimate connection with our existence. Remember, there is no substitute for water.

The one most essential characteristic of water is that it is dynamic. Water constantly evaporates from sea, lakes, and soil and transpires from foliage; is transported through the atmosphere; falls to the Earth; runs across the land; and filters down to flow along rock strata into aquifers. Eventually water finds its way to the sea again—indeed, water never stops moving.

A thought that might not have occurred to most people as they look at our glass of water is, "Who has tasted this same water before us?" Before us? Absolutely. Remember, water is a finite entity. What we have now is what we have had in the past. The same water consumed by Cleopatra, Jesus, Aristotle, Leonardo da Vinci, Napoleon, Joan of Arc, Donald Trump (and several billion other folks who preceded us), we are drinking now—because water is dynamic (never at rest) and because water constantly cycles and recycles, as discussed in another section.

Water never goes away, disappears, or vanishes; it always returns in one form or another.

OPENING THE FLOOD GATE

The availability of a water supply adequate in terms of both quantity and quality is essential to our very existence. One thing is certain: History has shown that the provision of an adequate quantity of quality potable water has been a matter of major concern since the beginning of civilization.

Water—especially clean, safe water—we know we need it to survive—we know a lot about it—however, the more we know, the more we discover we don't know—which seems to be the case in a lot of areas today.

Modern technology has allowed us to tap potable water supplies and to design and construct elaborate water distribution systems. Moreover, we have developed technology to treat used water (wastewater); that is, water we foul, soil, pollute, discard, and flush away.

Have you ever wondered where the water goes when you flush the toilet? Probably not.

An entire technology has developed around treating water and wastewater. Along with technology, of course, technological experts have been developed. These experts range from environmental/structural/civil engineers to environmental scientists, geologists, hydrologists, chemists, biologists, and others.

Along with those who design and construct water/wastewater treatment works, there is a large cadre of specialized technicians spread worldwide who operate water and wastewater treatment plants. These operators are tasked, obviously, with either providing a water product that is both safe and palatable for consumption and/or with treating (cleaning) a waste stream before it is returned to its receiving body (usually a river or stream). It is important to point out that not only are water practitioners who treat potable and used water streams responsible for ensuring quality, quantity, and reuse of their product, they are also tasked with, because of the events of 9/11, protecting this essential resource from terrorist acts.

The fact that most water practitioners know more about water than the rest of us comes as no surprise. For the average person, knowledge of water usually extends to knowing no more than that water is good or bad; it is terrible tasting, simply great, wonderful, clean, cool, and sparkling, or full of scum/dirt/rust/chemicals, great for the skin or hair, very medicinal, helps to clean out the digestive track and so on. Thus, to say the water "experts" know more about water than the average person is probably an accurate statement.

At this point, the reader is asking: What does all this have to do with anything? What does all this have to do with climate change and migration? Good questions.

What it has to do with water is quite simple. We need to accept the fact that we simply do not know what we do not know about water.

As a case in point, consider this: Have you ever tried to find a text that deals exclusively and extensively with the science of water? Such texts are few, far-flung, imaginary, non-existent—there is a huge gap out there.

Then the question shifts to—why would you want to know anything about water in the first place? Another good question.

To start with, let's talk a little about the way in which we view water.

Earlier, brief mention was made about the water contents of a glass of simple drinking water. Let's face it, drinking a glass of water is something that normally takes little effort and even less thought. The trouble is our view of water, and its importance is relative.

Relative? Well, the situation could be different—even more relative, however. For example, consider the young woman who is an adventurer; an outdoor-type person. She likes to jump into her four-wheel-drive vehicle and head out for new adventure. On this particular day she decided to drive through Death Valley, California—one end to another and back on a seldom-used dirt road. She had done this a few times before. During her transit of this isolated region, she decided to take a side road that seemed to lead to the mountains to her right.

She traveled along this isolated, hardpan road for approximately 50 miles—then the motor in her four-wheel-drive vehicle quit. No matter what she did, the vehicle would not start. Eventually, the vehicle's battery died; she had cranked on it too much.

Realizing that the vehicle was not going to start, she also realized that she was alone deep in an inhospitable area. What she did not know was that the nearest human being was about 60 miles to the west.

She had another problem—a problem more pressing than any other. She did not have a canteen or container of water—an oversight on her part—something not uncommon to many of us. Obviously, she told herself, this was not a good situation.

What an understatement this turned out to be.

Just before noon, on foot, she started back down the same road she had traveled. She reasoned she did not know what was in any other direction other than the one she had just traversed. She also knew the end of this side road intersected the major highway that bisected Death Valley. She could flag down a car, truck, or bus; she would get help, she reasoned.

She walked—and walked—and walked some more. "Gee, if it wasn't so darn hot," she muttered to herself, to sagebrush, to scorpions, to rattlesnakes and to cacti. The point is: it was hot, about 107°F.

She continued for hours, but now she was not really walking; instead, she was forcing her body to move along. The heat began playing games with her mind and her reasoning. She recalled the tale about Tantalus the Greek visitor to the underworld; he was there because of his doings: cannibalism, human sacrifice, and infanticide. Tantalus's punishment for his acts, now a proverbial term for temptation without satisfaction (the source of the English word *tantalize*), was to stand in a pool of water beneath a fruit tree with low branches. Whenever he bent down to get a drink, the water receded before he could get any. This fate cursed him with eternal deprivation of quenching his intolerable thirst.

She looked for fruit trees with low branches and imagined she was standing in a pool of water. Like Tantalus, she kept reaching down but she was able to reach and grab handfuls, but of sand and rocks only.

She was thirsty, almost mindless, and in terrible need of nourishment—she breathed only hot dryness, no relief there.

Each step hurt. She was burning up. She was thirsty. How thirsty was she? Well, right about now just anything liquid would do, thank you very much!

Later that night, after hours of walking through that hostile land, she couldn't go on. Deep down in her heat-stressed mind, she knew she was in serious trouble. Trouble of the life-threatening variety.

Just before passing out, she used her last ounce of energy to issue a dry pathetic scream.

This scream of lost hope and imminent death was heard—but only by the sagebrush, the scorpions, the rattlesnakes, and the cacti—and by the vultures that were circling above her parched, dead remains. The vultures were of no help, of course. They had heard these screams before. They were indifferent; they had all the water they needed; their food supply wasn't all that bad either.

This case sheds light on a completely different view of water. Actually, it is a basic view that holds we cannot live without it.

So, again, what does this have to do with climate change and water?

HISTORICAL PERSPECTIVE

An early human, wandering alone from place to place, hunting and gathering to subsist, would have had little difficulty in obtaining drinking water, because such a person would—and could—only survive in an area where drinking water was available with little travail. Little travail, that is, if they did not forget that the cave bear, saber-toothed tiger, and wolf drank the same water, and many drank it from the same place as early humans.

The search for clean, fresh, and palatable water has been a human priority from the very beginning. The author takes no risk in stating that when humans first walked the Earth, many of the steps they took were in the direction of water.

When early humans were alone or in small numbers, finding drinking water was a constant priority, to be sure, but it is difficult for us to imagine today just how big a priority finding drinking water became as the number of humans proliferated.

Eventually communities formed, and with their formation came the increasing need to find clean, fresh, and palatable drinking water and to find a means of delivering it from the source to the point of use.

Archeological digs are replete with the remains of ancient water systems (man's early attempts to satisfy that never-ending priority). Those digs (spanning the history of the last twenty or more centuries) testify to this. For well over 2,000 years, piped water supply systems have been in existence. Whether the pipes were fashioned from logs or clay or carved from stone or other materials is not the point—the point is, they were fashioned to serve a vital purpose, one universal to the community needs of all humans: to deliver clean, fresh, and palatable water to where it was needed.

These early systems were not arcane. Today, we readily understand their intended purpose. As we might expect, they could be crude, but they were effective, though they lacked in two general areas we take for granted today.

First, of course, they were not pressurized, but instead relied on gravity flow, since the means to pressurize the mains was not known at the time—and even if such pressurized systems were known, they certainly would not have been used to pressurize water delivered via hollowed-out logs and clay pipes.

The second thing that early civilizations lacked that we do not suffer from today (that is, in the industrialized world) is sanitation. Remember, to know the need for a solution exists (in this case, the ability to sanitize, to disinfect water supplies), first, the nature of the problem must be defined. Not until the middle of the 1800s (after countless millions of deaths from waterborne disease over the centuries) did people realize that a direct connection between contaminated drinking water and disease existed. At that point, sanitation of water supply became an issue.

When the relationship between waterborne diseases and the consumption of drinking water was established, evolving scientific discoveries led the way toward the development of technology for processing and disinfection. Drinking water standards were developed by health authorities, scientists, and sanitary engineers.

With the current lofty state of effective technology that we in the United States and the rest of the developed world enjoy today, we could sit on our laurels, so to speak, and assume that because of the discoveries developed over time (and at the cost of countless people who died [and still die] from waterborne-diseases), all is well with us—that problems related to providing us with clean, fresh, and palatable drinking water are problems of the past.

Are they really problems of the past? Have we solved all the problems related to ensuring that our drinking water supply provides us with clean, fresh, and palatable drinking water? Is the water delivered to our tap as clean, fresh, and palatable as we think it is? As we hope it is? Does anyone really know—for sure?

What we do know is that we have made progress. We have come a long way from the days of gravity flow water delivered via mains of logs and clay or stone. . . . Many of us on this earth have come a long way from the days of cholera epidemics.

However, to obtain a definitive answer to those questions, perhaps we should ask those who boiled their water for weeks on end in Sydney, Australia, in the fall of 1998. Or better yet, we should speak with those who drank the "city water" in Milwaukee in 1993 or in Las Vegas, Nevada—those who suffered and survived the onslaught of *Cryptosporidium* from contaminated water out of their tap.

Or if we could, we should ask these same questions of a little boy named Robbie, who died of acute lymphatic leukemia, the probable cause of which is far less understandable to us: toxic industrial chemicals, unknowingly delivered to him via his local water supply.

If water is so precious, so necessary for sustaining life, then two questions arise: (1) Why do we ignore water? (2) Why do we abuse it (pollute or waste it)?

We ignore water because it is so common, so accessible, so available, so unexceptional (unless you are lost in the desert without a supply of it). Again, why do we pollute and waste water? There are several reasons. Many will be discussed later in this text.

You might be asking yourself: Is water pollution really that big of a deal? Simply stated, yes, it is. Humans have left their footprints (in the form of pollution) on the environment, including on our water sources. Humans have a bad habit of doing this. What it really comes down to is "out of sight, out of mind" thinking. When we abuse our natural resources in any manner, maybe we think to ourselves: "Why worry about it? Let someone else sort it all out."

As this text proceeds, it will lead you down a path strewn with abuse and disregard for our water supply—then all (except the water) will become clear. Hopefully, we will not have to wait until someone does sort it, and us, out. Because, with time and everything else, there might be a whole lot of sorting out going on.

Let us get back to that gap in knowledge dealing with the science of water. The science of water is defined as the study of the physical, chemical, and biological properties of water, water's relationship to the biotic and abiotic components of the environment, occurrence and movement of water on and beneath the surface of the Earth, and the fouling and treatment of water.

Finally, before moving on with the rest of the text, it should be pointed out the view held throughout this work is that water is special, strange, and different—and vital. This view is held for several reasons, but the most salient factor driving this view is the one that points to the fact that on this planet, *water, without any substitute, is life.*

WATER CAN BE A DOUBLE-EDGED SWORD

There are two edges to the water sword that can be a problem. Well, wait a minute; isn't a double-edged sword one that carries a beneficial edge and a negative edge? True. Think of a battle sword that has two edges, and they serve no purpose that is beneficial other than self-defense or attack. Anyway, think of the water sword, where on one side there might not be enough water: drought. On the other edge, there might be too much water: flooding. Extremes of flood and drought are real; they exist. When we have water, the right amount of water that is clean and safe to drink and to use for whatever purpose, we may have a Goldilocks situation: where our supply of and the quality of water are exactly right.

But currently (2022), more than 550 million people are experiencing shortages of renewable water. In 88 developing countries with more than 40 percent of the world's population, the problem has become a serious constraint on development. Note that the total number of people experiencing water shortages (via extrapolation—guessing) is projected to reach somewhere around 3 billion by 2025 (a range of 2.8 billion to 3.3 billion), largely due to population growth, and it could even grow to the order of 4.4 billion by 2050 (Engleman and Leroy, 1993; Spellman, 2021).

Keep in mind that water shortages cause major problems for health, irrigation, agriculture, and industry. As early as the 1990s, the World Health Organization stated that a full 90 percent of developing country disease is due to lack of clean water for domestic use, and four out of five deaths stem from water-related disease, especially diarrhea (World Health Organization, 1992).

Okay, let's talk about drought. What is drought? What are the causes of drought? Drought, in simple terms, is a natural and regularly occurring phenomenon common to many inhabited regions, including those not always associated with dryness, such as Olympic National Forest (Spellman, 2021). In even simpler terms, drought occurs when precipitation falls below expected levels for a period sufficiently long for adverse effects on natural or human systems to be noticed. Note that droughts affect more people than any other climate hazard (Wilhite et al., 2007). And the irony about droughts and their causes is that we do not know what we do not know about droughts—our information remains patchy.

Okay, we do not know everything about drought—so what do we know? Well, we know that water constantly cycles (the water or hydrological cycle) from water vapor to rain and snow falling onto soils and across and beneath the landscape. As climate change occurs, Earth's atmosphere warms due to greenhouse gases, and the satellite data record continues to get longer and more detailed, scientists are studying how climate change is affecting the distribution of water. Distribution of water, where it is located, is extremely important because you can have all the safe drinking water there is, but if it is not accessible, then there is a problem—a big one. Distribution of water around the globe is about knowing not only about changes in distribution but also where it rains and doesn't and how much and how frequently rain falls versus light rain. Note that rainfall amount impacts soil saturation and how high streams and rivers rise, which then changes their holding capacity in the event of another storm. The lack of rain stresses crops, vegetation, and supplemental water reserves, and when their frequency increases, those reserves are less likely to recover before the next dry spell.

Data show that trends are beginning to emerge, especially at the extremes, in the frequency and magnitude of floods and droughts. These trends are important because they affect everything from local weather to where crops can grow and have consequences that will ripple through communities today and in the coming decades—remember, global population growth is on the rise, and we simply can't survive without safe drinking water.

Data contained in the recent historical record show an increase in very heavy precipitation events across the United States from 1958 to 2016. These heavier rainfall events were most significant in the northeastern states, but also midwestern and even southeastern states experienced increases.

Remarkable consensus exists among worldwide climate and water experts over the current issues confronted by water managers and others. These issues include the following:

1. Water Availability, Requirements, and Use
   • Protection of aquatic and wetland habitats
   • Management of extreme events (droughts, floods, etc.)
   • Excessive extractions from surface and groundwater
   • Global climate change
   • Safe drinking water supply
   • Waterborne commerce
2. Water Quality
   • Coastal and ocean water quality
   • Lake and reservoir protection and restoration

- Water quality protection, including effective enforcement of legislation
- Management of point- and nonpoint-source pollution
- Impacts on land/water/air relationships
- Health risks

3. Water Management and Institutions
   - Coordination and consistency
   - Capturing a regional perspective
   - Respective roles of federal and state/provincial agencies
   - Respective roles of projects and programs
   - Economic development philosophy that should guide planning
   - Financing and cost sharing

4. Information and Education
   - Appropriate levels of regulation and deregulation
   - Water rights and permits
   - Infrastructure
   - Population growth
   - Water resources planning, including
     i.   Consideration of the watershed as an integrated system
     ii. Planning as a foundation for, not a reaction to, decision making
     iii. Establishment of dynamic planning processes incorporating periodic review and redirection
     iv. Sustainability of projects beyond construction and early operation
     v.   A more interactive interface between planners and the public
     vi. Identification of sources of conflict as an integral part of planning
     vii. Fairness, equity, and reciprocity between affected parties

Another significant issue (the other edge of the water sword) with precipitation is if it occurs in excess, there is the possibility of flooding.

A stream channel influences the shape of the valley floor through which it courses. The self-formed, self-adjusted flat area near the stream is the *flood plain*, which loosely describes the valley floor prone to periodic inundation during over-bank discharges. What is not commonly known is that valley flooding is a regular and natural behavior of the stream.

Without floodplains and meanders, the water moves more swiftly, and silt carried in the water is more likely to be swept to sea.

NOTE

1. The following information is based on F. Spellman's (2021) *The Science of Water*, 4th ed. Boca Raton, FL: CRC Press.

REFERENCES

Engleman, R., and Leroy, P. (1993). *Sustaining Water: Population and the Future of Renewable Water Supplies.* Washington, DC: Population Action International.
Lewis, S.A. (1996). *The Sierra Club Guide to Safe Drinking Water.* San Francisco, CA: Sierra Club Books.

Spellman, F.R. (2021). *Climate Change*. Lanham, MD: Bernan Press.

USGS. (2013). *Why Is the Ocean Salty?* Accessed 01/05/14 @ http://ga.water.usgs.gov/edu/whyoceansalty.html.

Wilhite, D.A., Svoboda, M.D., and Haye, M.J. (2007). Understanding the complex impacts of drought: A key to enhancing drought mitigation and preparedness. *Water Resour Manag.* 21:763–774.

World Health Organization. (1992). *Water Supply, Sanitation, and Health*. Geneva, Switzerland: WHO.

SUGGESTED READINGS

Postel, S., Daily, G.C., and Ehrlich, P.R. (1996). Human appropriation of renewable fresh water. *Science* 271:785–788.

3 When the Fields Blow in the Wind

BLACK BLIZZARDS

Renowned literary genius Charles Dickens stated in one of his classic works: "It was the best of times, it was the worst of times." This statement fits many times and situations in the history of humans, including times of conflict between family and love, hatred, and oppression, good and evil, light and darkness, and wisdom and folly. Truth be told, it is nearly impossible to write a classic compare-and-contrast work to match Dickens' classic tale. However, human events have a way of paralleling literature and vice versa. This certainly could be the case when we compare and contrast the twenties to the "Dirty Thirties" (a.k.a. the Dust Bowl era).

It is not outside the boundaries of reason to make the connection between the Dust Bowl of the thirties and the human migration it generated—the great human migration from America's High Plains to greener pastures out west. In his masterpiece *The Grapes of Wrath*, John Steinbeck gave voice to those who fled the Dust Bowl. Moreover, it is not outside the boundaries of reason to connect the migration from the Dust Bowl to climate: severe drought and a failure to apply dryland farming methods to prevent wind erosion (aeolian processes) that caused the occurrence (McLeman et al., 2014; Cook et al., 2018).

DID YOU KNOW?

Dryland farming encompasses specific agricultural techniques for the non-irrigated cultivation of crops. As the name implies, dryland farming is associated with drylands, areas characterized by a cool wet season followed by a warm dry season. Dryland farming has evolved as a set of techniques and management practices used by farmers to continually adapt to the lack of moisture in a given crop cycle.

Okay, it is definition time. The Dust Bowl and the "Dirty Thirties" are two terms that are not hieroglyphics requiring the Rosetta Stone to decipher. But for clarity (in science, clarity is boss—or should be, must be, has to be), Dust Bowl is a colloquial term, as is the term Dirty Thirties. So what does the colloquial "Dust Bowl" consist of; that is, what are its boundaries? Well, what is certain is that the black Sunday dust storm of April 14, 1935, drifted across Oklahoma, Texas, Kansas, and Colorado. Truth be told, the exact boundaries of the Dust Bowl are subjective and do not take up a large segment of the Great Plains. However, the worst or most severe events

DOI: 10.1201/9781003295211-3

37

between 1935 and 1938 occurred in the Oklahoma-Texas panhandles, the southwestern corner of Kansas, the northeast corner of New Mexico, and the southeast corner of Colorado. The events not listed as most severe but at the severe level occurred in 1935–1936 and 1938 in these same states but to a more extensive, wide-ranging degree, reaching southern Nebraska and deeper into Texas and New Mexico.

The Great Plains is an extensive semi-arid region 2,000 miles (3,200 km) in length, 500 miles (800 km) wide and 1,100,000 square miles (2,800,000 km²) in total area, stretching from southern Texas to beyond Moose Jaw, Saskatchewan, into parts of Alberta and Manitoba, Canada.

Research related to the Dust Bowl era focuses on global environmental change, drought modeling, atmospheric circulation, land management, institutional behavior, adaptation processes, and human migration (in this text better known as climate refugees). Study of the Dust Bowl era helps us gain an understanding of human migration drivers, that is, drivers that convert local inhabitants into climate refugees.

When someone, anyone, uses the old idiom "the icing on the cake," this is a term that many of us interpret to mean something that is an additional benefit or a positive aspect to something that is already considered positive or beneficial. The point is, you may not think that my choice of describing the Great Depression as "the icing on the cake of the Dust Bowl era" is appropriate. However, I use this terminology to paint a picture of both tragic events as being enhancements of the other; that is, I view the Great Depression as an enhancement of the Dust Bowl era and vice versa. During the 1930s, the two were synergistic or, simply stated, coactive.

This enhancement (coactivity) of the situation is not that far off base when you consider that during the worst years of the Great Depression, large areas of the North American Great Plains experienced uncompromising, multi-year droughts that led to soil erosion, dust storms, farm abandonments, personal hardships, dearth, scarcity, famine, and death (mostly caused by dust pneumonia or suicide). For those able to gravitate to greener pastures, these United States climate refugees became part of the distress migration on scales not previously seen—to this very day and time.

Okay, what about the causes and consequences of the tragic Dust Bowl events of the 1930s? In Chapter 2, we discussed water—when there is not enough to sustain life; when there is too much to float all boats, houses, schools, hospitals; you name it. In this chapter we are talking about drought . . . little to no water or water that is unfit for even a mosquito to indulge in.

Throughout the globe, drought and its reoccurring frequency are of increasing interest, not only to scientists but to those dying of thirst. To learn from the past, we must study the past and the lessons learned (hopefully) and do all we can to prevent such occurrences. Keep in mind that drought not only has ecological consequences but also socio-economic impacts. Note that after years of study, I have found that our knowledge of the physical causes and human impact of droughts and of the Dust Bowl era is incomplete, but one thing seems certain: the Dust Bowl era has much to teach us about climate change and its enduring implications.

So what exactly happened to make the 1930s the Dust Bowl era in certain parts of the Great Plains region of the United States? Earlier it was mentioned that drought and improper farming methods caused the great dust storms that destroyed the wheat and other crops and drove many of the residents to leave their homes and migrate to the west (many of these climate migrants ended up in California). When local farmers lost their farms due to the drought, dust storms, or bank foreclosure, they abandoned the land. Note that the massive migration from the Dust Bowl region did not include all the residents within the Dust Bowl; three out of four farmers stayed in their homes and retained their land. Even though three-quarters of the population remained in the Dust Bowl region, the mass migration depleted the population in certain areas.

The Dust Bowl migration was the largest in American history. By 1940, more than 2 million people had moved out of the Dust Bowl states. Of those, approximately 200,000 moved to California. In the realist novel *The Grapes of Wrath*, Steinbeck focused on the Joads, a poor family of tenant farmers driven from their Oklahoma home by drought, economic hardship, agricultural industry changes, and bank fore-closures that forced tenant farmers out of work. Because of their desperate situation and being trapped in the Dust Bowl, Tom Joad leads his family out of Oklahoma, using Route 66 to go to California. Why did the Joads and thousands of other Dust Bowl residents head for California? Thousands of Dust Bowl residents thought of that golden state as a paradise of sorts—certainly better than the Dust Bowl. The climate migrants were seeking jobs, land, self-esteem, and a future.

The stark reality and the irony of this account is that the migrants who fled to a life in California found it as difficult as the one they had left.

BLACK SUNDAY, APRIL 14, 1935

To this point in the book, a fairly adequate description of the Dust Bowl has been given. But to really gain knowledge of something (anything), it is important to get different views on topics, including historic happenings. Thus, I have chosen the U.S. National Weather Service's (NWS) 01/31/2022 view to relate here.

As already pointed out, the 1930s were times of tremendous hardship on the Great Plains. The settlers dealt with the double whammy—Great Depression and years of drought that plunged an already suffering society into an onslaught of relentless dust storms for days, weeks, and months on end. They were known as dirt storms, sand-storms, "dusters," and black blizzards. It seemed as if it could get no worse, but on Sunday, the 14th of April 1935, it got worse—in magnitudes beyond description. The day is known in history as "Black Sunday," when the wind "scoop[ed] the dust like hands" and sent mountain of blackness sweeping across the high plains, instantly turning a warm, sunny afternoon into a horrible blackness that was darker than the darkest night. Famous songs were written about it, and on the following day, the world would hear the region referred to for the first time as "The Dust Bowl" (NWS, 2022).

At first it was a wall of blowing sand that blasted its way into and through the eastern Oklahoma panhandle around 4 PM. Then it raced to the south and southeast across the main body of Oklahoma that evening, accompanied by heavy blowing dust, winds of 40 MPH or more, and rapidly falling temperatures. But the heavy hammers, the worst conditions, were in the panhandle, where the rolling mass raced toward the southwest—accompanied by a massive wall of blowing dust that resem-bled a land-based tsunami. Winds in the panhandle reached upwards of 60 MPH, and for at least a brief time, the blackness was so complete that one could not see their own hand in front of their face (NWS, 2022).

The blowing dust that blasted the High Plains was attributed to dry soil conditions but also to poor soil conservation techniques that were in use during the 1930s. It was one of President Roosevelt's advisors, Hugh Hammond Bennett, who sounded the alarm bell before Congress about the poor soil conservation techniques. The irony is that Bennett testified weeks before Black Sunday about the soil conservation problem, and it was the dust cloud that reached the Capital and blotted out the sun that convinced Congress to pass the Soil Conservation Act.

THE GREAT PLAINS COMMITTEE

The U.S. government established the Great Plains Committee in the 1930s to identify the causes, impacts, and necessary remedies for the crisis in the region. Table 3.1 summarizes the key findings—which put much of the blame on land settlement patterns and land use practices of the late nineteenth century—and the recommendations for action.

TABLE 3.1
Summary of Key Findings of the Great Plains Committee (1936)

Outcome	Root Causes	Suggested Farm-Level Action
Soil erosion	High rates of farm tenancy and absentee landlords mean over-production of crops relative to livestock; soil mining; lack of farm improvement/long-term planning; expansion of farming into marginal areas; over-cultivation of small landholdings; failure to recognize diversity of soil conditions across the region	Plow along contours; list and furrow fields at right angles to prevailing winds; plant crops in strips; terrace slopes; till soil roughly and leave high stubble after harvest; avoid bare summer fallow in wind-exposed areas and instead rotate in cover crops like clover; plant windbreaks
Loss of forage cover for grazing	Overstocking of range lands; expansion of farming into marginal areas	Reduce herd sizes or keep herds off fragile lands
Inefficient use of water	Poor farming technologies and practices fail to conserve soil moisture; inadequate capacity for irrigation	Create deeper, better water ponds for livestock; use supplemental irrigation where cost effective to do so
Highly variable farm incomes; high rates of farm indebtedness	Undue dependence on wheat as a cash crop; high rates of tenancy; family farm landholdings too small in size; mechanization in 1920s financed on credit during period of good rainfall and favorable crop prices	Maintain a higher ratio of fodder and livestock to cash crops; reduce proportion of wheat and corn on farms; create diversified operational plans; keep larger feed and seed reserves

Source: Adaptation from McLeman, R., Dupre, J., Berang Ford, L., Ford, J., Gajewski., K., Marchildon, G. (2014). *What We Learned From the Dust Bowl: Lessons in Science, Policy, and Adaptation.* Accessed 01/30/2022 @ www.ncbi.nim.nih.gov/pmc/articles/PMC4015056.

DID YOU KNOW?

Those who lived through the Dust Bowl, that is, not only those who migrated to other locations but also those who stayed put, were exposed to the dust in a manner whereby they inhaled it. Inhalation of dust can lead to dust pneumonia. This form of respiratory disorder affected a substantial number of people during the 1930s Dust Bowl era.

THE BOTTOM LINE

When migrants from the Dust Bowl states arrived in California, they found themselves in unfamiliar territory for more than one reason. First of all, many of the California farms were corporate owned. They were large and more modernized than anything the migrants were used to or had ever seen before, and the crops were unfamiliar. The rolling fields of wheat were replaced by crops of fruit, nuts, and vegetables. In Steinbeck's classic, *The Grapes of Wrath*, he pointed out that about 40 percent of the migrant farmers wound up in the San Joaquin Valley, picking grapes and cotton. The migrants replaced the Mexican migrant workers—a legion

FIGURE 3.1   Dust Bowl children.

Source: Dust Bowl exodus. Public domain photo. Photo by Dorothea Lange. Accessed 05/08/2022 @ https://picryl.com/media/children-of-oklahoma-drought-refugee-in-migratory-camp-in-california.

of workers numbering around 120,000. To say the life of the migrants was hard is to make a gross understatement. They were paid by the quantity of fruit and cotton picked and earned seventy-five cents to $1.25 a day. And out of that, they had to pay for their board and room.

All the words in the world can't accurately describe the horrors of the Dust Bowl and the migrants. However, Figure 3.1 might relay the horror—a picture is worth a thousand words, they say, and this is.

REFERENCES

Cook, B., Miller, R., and Seager, R. (2018). Did dust storms make the Dust Bowl worse? Accessed 01/30/2022 @ http://ocp.ideo.columbia.edu/res/res/div/ocp/drought/dust_storms.shtml.

McLeman, R., Dupre, J., Berang Ford, L., Ford, J., Gajewski, K., and Marchildon, G. (2014). What we learned from the Dust Bowl: Lessons in science, policy, and adaptation. Accessed 01/30/2022 @ www.ncbi.nim.nih.gov/pmc/articles/PMC4015056.

National Weather Service (NWS). (2022). The Black Sunday dust storm of April 14, 1935. Accessed 02/01/2022 @ www.weather.gov/oun/events-19350414.

4 When Opportunity Vanishes

NO WATER, NO FOOD, NO JOB . . . NO NOTHING!

When you have no water to quench your thirst, or when the water available will make you sick or worse, or when you have nothing to breathe but the air you can see that chokes you or worse, maybe it is time to get out of Dodge and find those greener pastures that are just beyond your reach but may be attainable if you migrate.

There are several drivers of migration. Drivers are the factors that get migration going and keep it going once begun. We have discussed the water and dust storms or agricultural failure drivers. In this chapter, we discuss another powerful driver of migration: opportunity. While it is often assumed that it is really poverty that drives migrants to migrate, it must be said that if one is poor, truly poor, it is difficult (if not impossible) to obtain the resources needed to make a journey from one's place of origin to the United States, for example. I did find, in some cases, that poverty was a powerful driver for migration for a few migrants that I interviewed in early 2021 in Tucson, Arizona, although there are many of my associates who do not share this view because they are missing some facts or factors. For example, as stated, an argument can be made that in order to migrate, one must have resources. There is considerable debate about lack of resources holding back migrants. Those currently migrating from Central America to the United States are said to have little to no resources available to them when they leave their original locations and travel thousands of miles to get to the open southern border to enter the United States. However, the Coyotes who facilitate the migrants' passage through Mexico and other locations do not provide their services gratis; they collect whatever funds they can from the migrants.

You might ask or wonder about the migrants fleeing from a life of poverty to get to "The Land of Opportunity." How and where do poverty-driven migrants get the financial resources to pay those who facilitate their travel? It's the old refrain "Where's the money?" In speaking with a couple of migrants from Central America while I was in Tucson, Arizona, in early 2021, I asked this question more than once, and I received some interesting replies. Most replied that they had saved enough money to make the trip, and family members pooled their resources and sold whatever belongings they could turn into money. At first this reply, and a few others, puzzled me. Why? Well, I reasoned that if the migrants were able to produce a few thousand dollars to make the journey north, then their circumstances within the native country could not have been all that bad. Or so it seemed to me. The ability to save enough money for their journey simply seemed to indicate that the pressure to migrate was or had been reduced. But developmental changes within their native countries only provided the

DOI: 10.1201/9781003295211-4

funds to leave and not the desire to stay. I also found out that some of the migrants had made "a deal" or "trato" for mucho dinero with the Coyotes or other enablers. One middle-aged migrant told me that there were other ways in which to gain favor with the Coyotes, especially for the female migrants (I did not pursue an explanation of what this meant—seemed obvious). Anyway, one of the "deals" with the Coyotes that seemed commonly mentioned by the migrants was that they had agreed to pay the Coyotes a certain amount of dollars once the migrants became established in the United States and were gainfully employed and making mucho dinero. Note that I attempted to question a few other migrants, but I did not have an interpreter available to translate Chinese, Portuguese, and other languages of which I had no understanding or those of where the migrant was from: his or her place of origin.

So the question the reader may have at this point and time is what does poverty have to do with climate migration? There are several factors that contribute to a migrant's decision to leave his or her native country. Note again that this book is not a political book—meaning political, social, and cultural migration drivers are not the focus here. Instead, the focus is on the climate change drivers where environmental change, marked environmental degradation, and/or environmental disasters (natural disasters) are the causes that drive migrants to migrate. These causes, factors, or precipitators of climate change migration that have contributed to poverty in migrants' places of origin include an absence of resources, poor soil fertility, lack of safe drinking water, lack of forest cover, major climate change, and other climate-related factors. Poverty factors based on climate change are those that affect job opportunities (lack thereof), impact income (too low to sustain lifestyle), and/or have an economic impact on producer/consumer prices.

While climate change is a factor that may be a driver of poverty, it should be pointed out that the lack of opportunity is a more salient reason for inhabitants of their place of origin to migrate to locations such as the United States. I say this because when I interviewed English- and Spanish-speaking migrants in Tucson, many of them made the point that it was indeed opportunity that drove them to leave their places of origin. Several migrants stated that environmental conditions at their place of origin, including forest fires, lack of water and food, and severe weather events, made life there intolerable for them. And the kicker, the bottom-line driving factor, was enforced via their communication with relatives or friends already living in the United States who made them realize that opportunity in the United States was, and is, better than in their place of origin. This communication was made available to the migrants via Internet, phone, and text messaging when communicating with those in the United States living "how the other half lives"—with a better life and plenty of opportunity; this became their main driver: OPPORTUNITY—it still is.

5 When Disease Is a Passenger

CLIMATE CHANGE INCREASING RISKS OF DISEASE

As the nation's (United States) public health leader, the Centers for Disease Control and Prevention (CDC) is actively engaged in a national effort to protect the public's health from the harmful effects of climate change. Scientists from the CDC's National Center for Emerging and Zoonotic Infectious Diseases (NCEZID) are at the forefront of many of these efforts.

Lyme disease, West Nile virus disease, and valley fever are just some of the infectious diseases that are on the rise and spreading to areas of the United States. Warmer weather, especially summers, and fewer days of frost make it easier for these and other infectious diseases to expand into new geographic areas and to infect more people. On a global scale, climate change driven by anthropogenic (a.k.a. human-caused) greenhouse gas emissions is particularly felt in Canada, with warming greater there than in the rest of the world. Continued warming will be accompanied by increases and decreases in precipitation, increasing climate variability, and extreme weather events. Climate change is likely to drive migration and the emergence of infectious disease on a global scale. However, with migration increasing in the United States and Canada, disease is likely to accompany the migrants. To understand climate change's impact, it is important to look at some of the ways these diseases spread—through mosquito and tick bites, contact with animals, fungi, and water.

Climate change increases the number and geographic range of disease-carrying migrants, animals, insects, and ticks. Climate change poses many risks not only to migrants' health but to residents of the United States and other countries. In the United States, health impacts of climate change are already being felt. We need to safeguard our communities by protecting people's health, well-being, and quality of life from climate change impacts and from those unwanted tag-along diseases carried by migrants as they enter the United States. We must place increased emphasis on guarding against diseased or disease-carrying migrants (including disease-carrying animal and insect migrants); this effort should be as important as ensuring we protect against migrants who are terrorists or criminals of all types.

VECTORBORNE DISEASE

One way climate change could affect human health is by increasing the risk of vector-borne diseases. A vector is any organism—such as fleas, ticks, or mosquitoes—that can transmit a pathogen, or infectious agent, from one host to another. Longer warm seasons, earlier spring seasons, shorter and milder winters, and hotter summers are

DOI: 10.1201/9781003295211-5

conditions that could become more hospitable for many carriers of vectorborne diseases. It is important to note that if climate change is considered a driver of migration, then it is plausible or reasonable to assume that human migrants can and will bring along unwanted passengers in the form of disease. Moreover, the animals and insects that carry disease along with them also migrate to more comfortable surroundings, and this is a prominent issue in maintaining good health for humans and other residents of the ecosystem into which they migrate.

TICK AND MOSQUITO BITES

Mild winters, early springs, and warmer temperatures are giving ticks and mosquitoes more time to reproduce, spread diseases, infect migrants, and expand their habitats throughout the United States. Between 2004 and 2018, the number of reported illnesses from tick, mosquito, and flea bites more than doubled, with almost 800,000 cases reported in the United States (CDC, 2021).

During this period nine new germs spread by mosquitoes (see Figures 5.1–5.3) and ticks (see Figures 5.4–5.5) were discovered or introduced or carried into the United States. The geographic ranges where ticks spread Lyme disease, anaplasmosis, ehrlichiosis, and spotted fever rickettsiosis have expanded, and experts predict tickborne diseases will continue to increase and possibly worsen. Longer, warmer summers have also given mosquitoes more time to reproduce and spread diseases. In 2012, a mild winter, early spring, and sweltering summer set the stage for an outbreak of West Nile virus in the United States, resulting in more than 5,600 illnesses and 286 deaths.

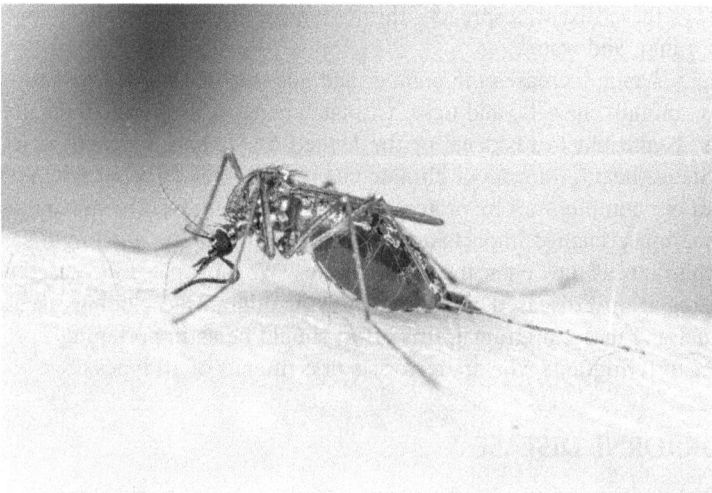

FIGURE 5.1   Photo of an adult female *Aedes aegypti* mosquito biting a person.

Source: CDC (2022) public domain image library. Accessed 02/03/2022 @ www.cdc.gov/mosquitoes/gallery/index.html.

FIGURE 5.2 Photo of *Culex* species mosquito larvae in standing water.

Source: CDC (2022) public domain image library. Accessed 02/03/2022 @ www.cdc.gov/mosquitoes/gallery/index.html.

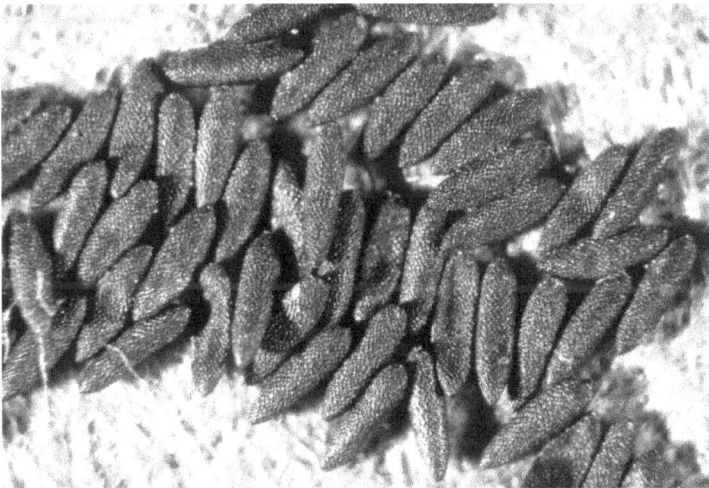

FIGURE 5.3 Photo of *Aedes triseriatus* mosquito eggs.

Source: CDC (2022). Public domain image library. Accessed 02/03/2022 @ www.cdc.gov/mosquitoes/gallery/index.html.

FIGURE 5.4   Photo of an adult female Rocky Mountain wood tick, *Dermacentor andresoni*, on a blade of grass.

Source: CDC (2022). Public domain image library. Accessed 02/03/2022 @ www.cdc.gov/ticks/gallery/index.html.

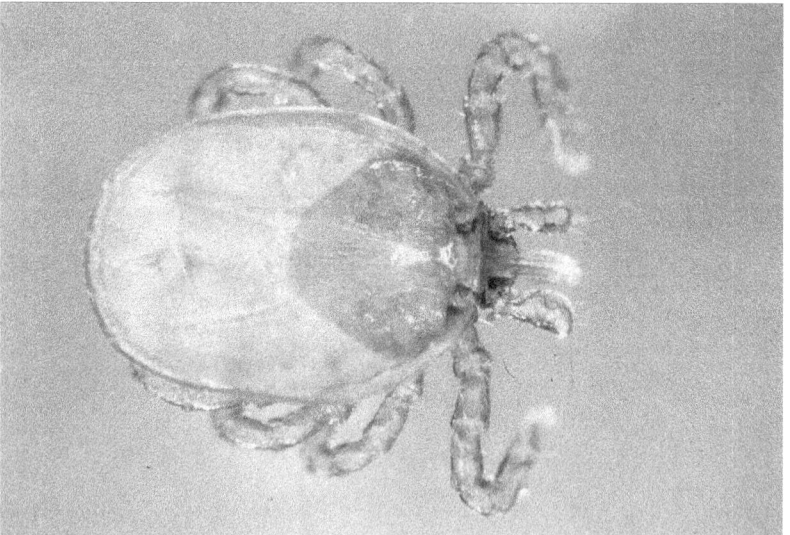

FIGURE 5.5   Photo of groundhog tick, *Ixodes cookei.*

Source: CDC (2022). Public domain image library. Accessed 02/03/2022 @ www.cdc.gov/ticks/gallery/index.html.

CONTACT WITH ANIMAL MIGRANTS

Climate change has forced some animal species to migrate into new habitats as their natural habitats disappear, and it has expanded the habitats of other animals. This movement of animals into new areas increases opportunities for contact between humans and animals and the potential spread of zoonotic diseases, as these examples show:

- Wildlife carrying the rabies virus are expanding into new geographic areas of the country. Note that rabies only affects mammals. Any mammal can get rabies, including people. While rabies is rare in people in the United States, with only one to three cases reported annually, about 60,000 Americans get post-exposure prophylaxis (PEP) each year to prevent rabies infection after being bitten or scratched by an infected or suspected infected animal. In the United States, more than 90 percent of reported cases of rabies in animals occur in wildlife—from raccoons, skunks, bats, and foxes. Note that contact with infected bats is the leading cause of human rabies deaths in this country; at least seven out of ten Americans who die from rabies in the US were infected by bats. The problem with bat bites is that they are generally small and difficult for people to identify, but these contacts can still spread rabies (CDC, 2022).
- As global temperatures rise, deadly diseases that are a threat in other countries—like Ebola, Lassa, Rift Valley fever, and monkeypox—will increase along with the risk of them being imported (via migration and migrants) into the United States.
- Arctic temperatures are rising more than twice as rapidly as in the rest of the world. Warming temperatures in Alaska have led to increases in vole populations, which can spread diseases like Alaskapox (a species of orthopoxvirus that causes skin lesions in humans). Additionally, a recent report by the National Park Service and University of Maryland described two unexpected drivers for migration timing that provide insight into potential future effects of climate change on caribou. First, the researchers found that caribou herds across North America are triggered to start migration at the same time by large-scale, ocean-driven climate cycles. Second, despite an in-step start, arrival at their respective calving grounds depends on the previous summer's weather conditions. They found that warm windless summers that favored insect pests led to poorer maternal health and delayed arrivals at the calving grounds the following spring (Gurarie et al., (2019).

FUNGI

Rising temperatures have allowed certain disease-causing fungi to migrate, to spread into new areas that previously were too cold for them to survive. For example, valley fever—caused by a fungus (coccidioides; see Figure 5.6) that lives in soil in hot dry areas—has already spread into the Pacific Northwest. This fungus can cause severe

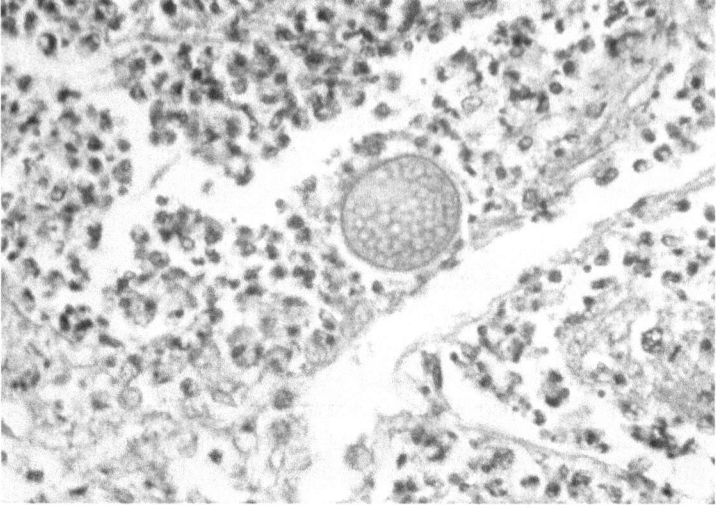

FIGURE 5.6   Coccidioidomycosis of lung.

Source: CDC by Dr. Lucille K. Georg (2012). Public Health Image Library. PublicDomainFiles.com. Accessed 05/08/2022 @ www.publicdomainfiles.com/show_file.php?id=13527357015912.

infections and death and is often misdiagnosed and treated inappropriately. As the difference between environmental temperatures and human body temperatures narrows, new fungal diseases may emerge as fungi become more adapted to surviving in humans. Climate change also increases the risk for natural disasters and flooding, which can increase the risk for mold to grow in people's homes. Certain molds can cause deadly infections of the lungs and brain.

WATER

Scientists predict that climate change will have devastating effects on freshwater and marine environments. For example, we could see more frequent and more severe instances of harmful alga blooms, which are the rapid growth of algae or cyanobacteria in lakes, rivers, oceans, and bays. Warming temperatures in Lake Erie have contributed to extensive toxic blooms that last into the early winter months. Harmful algal blooms can look like foam, scum, paint, or mats on the surface of water and can be assorted colors (see Figure 5.7). They endanger our health when we eat contaminated shellfish.

Cyanobacteria, formerly classified as blue-green algae, are autotrophic organisms that are able to synthesize organic compounds using CO_2 as the major carbon source. Cyanobacteria produce O_2 as a by-product of photosynthesis, providing an O_2 source for other organisms in the ponds. They are found in exceptionally large numbers as blooms when environmental conditions are suitable. Commonly encountered cyanobacteria include *Oscillatoria, Arthrospira, Spirulina*, and *Microcystis* (Vasconcelos and Pereira, 2001; Spellman, 2000).

Algal blooms can harm pets, livestock, wildlife, and the environment. While no human deaths caused by cyanobacteria have been reported in the United States, some of the toxins can make dogs and other animals sick and possibly even cause death

FIGURE 5.7   Photo of cyanobacteria.

Source: ASU Department of Energy 2014. Photo by Quentin Kruger. PublicDomainFiles.com. Accessed 05/08/2022 @ www.publicdomainfiles.com/show_file.php?id=14022145818779.

within hours to days. Dog deaths have been reported after dogs swam in or drank freshwater containing cyanobacterial toxins.

With regard to the environment, one of the major concerns is over the appearance of dead zones in various water bodies (i.e., excess nutrients cause oxygen-consuming algae to grow and thus create oxygen-deficient dead zones). For example, in recent years it has not been uncommon to find several dead zones in the Chesapeake Bay region; consider the following case study.

Case Study 1.1   Chesapeake Bay Cleanup

The following newspaper article, written by the author, appeared in the January 5, 2005, edition of *The Virginian-Pilot*. It is an op-ed rebuttal to the article referenced in the following text. It should be pointed out that this article was well received by many, but a few stated that it was nothing more than a rhetorical straw man. Of course, in contrast, the organizational critics were using the rhetorical Tin Man approach. That is, when you need to justify your cause and your organization's existence and need more grease, you squawk. The grease that many of these organizations require, however, is grease the consistency of paper-cloth and is colored green—thus, they squawk quite often. You be the judge.

* * * *

CHESAPEAKE BAY CLEANUP: GOOD SCIENCE VS. "FEEL GOOD" SCIENCE

In your article, "Fee to help Bay faces anti-tax mood" (Va. Pilot, 1/2/05), you pointed out that environmentalists call it the "Virginia Clean Streams Law." Others call it a "flush tax." I call the environmentalist's (and others) view on this topic a rush to judgment, based on "feel good" science vs. good science. The environmentalists should know better.

Consider the following:

Environmental policymakers in the Commonwealth of Virginia produced what is called the Lower James River Tributary Strategy on the subject of nitrogen (a nutrient) from the Lower James River and other tributaries contaminating the Lower Chesapeake Bay Region. When in excess, nitrogen is a pollutant. Some "theorists" jumped on nitrogen as being the cause of a decrease in the oyster population in the Lower Chesapeake Bay Region. Oysters are important to the local region. They are important for economical and other reasons. From an environmental point of view, oysters are important to the Lower Chesapeake Bay Region because they have worked to maintain relatively clean Bay water in the past. Oysters are filter-feeders. They suck in water and its accompanying nutrients and other substances. The oyster sorts out the ingredients in the water and uses those nutrients it needs to sustain its life. Impurities (pollutants) are aggregated into a sort of ball that is excreted by the oyster back into the James River.

You must understand that there was a time, not all that long ago (maybe 50 years ago) when oysters thrived in the Lower Chesapeake Bay. Because they were so abundant, these filter-feeders were able to take in turbid Bay water and turn it almost clear in a matter of three days. (How could anyone dredge up, clean, and then eat such a wonderful natural vacuum cleaner?)

Of course, this is not the case today. The oysters are almost all gone. Where did they go? Who knows?

The point is that they are no longer thriving, no longer colonizing the Lower Chesapeake Bay Region in numbers they did in the past. Thus, they are no longer providing economic stability to watermen; moreover, they are no longer cleaning the Bay.

Ah! But don't panic! The culprit is at hand; it has been identified. The "environmentalists" know the answer—they say it has to be nutrient contamination; namely, nitrogen is the culprit. Right?

Not so fast.

A local sanitation district and a university in the Lower Chesapeake Bay region formed a study group to study this problem formally, professionally, and scientifically. Over a 5-year period, using Biological Nutrient Removal (BNR) techniques at a local wastewater treatment facility, it was determined that the effluent leaving the treatment plant and entering the Lower James River consistently contained below 8 mg/L nitrogen (a relatively small amount) for 5 consecutive years.

The first question is: Has the water in the Chesapeake Bay become cleaner, clearer because of the reduced nitrogen levels leaving the treatment plant?

The second question is: Have the oysters returned?

Answer to both questions, respectively: no; not really.

Wait a minute. The environmentalists, the regulators, and other well-meaning interlopers stated that the problem was nitrogen. If nitrogen levels have been reduced in the Lower James River, shouldn't the oysters start thriving, colonizing, and cleaning the Lower Chesapeake Bay again?

You might think so, but they are not. It is true that the nitrogen level in the wastewater effluent was significantly lowered through treatment. It is also true that a major point source contributor of nitrogen was reduced with a corresponding decrease in the nitrogen level in the Lower Chesapeake Bay.

If the nitrogen level has decreased, then where are the oysters?

A more important question is: What is the real problem?

The truth is that no one at this point and time can give a definitive answer to this question.

Back to the original question: Why has the oyster population decreased?

One theory states that because the tributaries feeding the Lower Chesapeake Bay (including the James River) carry megatons of sediments into the bay (stormwater runoff, etc.) they are adding to the Bay's turbidity problem. When waters are highly turbid, oysters do the best they can to filter out the sediments but eventually they decrease in numbers and then fade into the abyss.

Is this the answer? That is, is the problem with the Lower Chesapeake Bay and its oyster population related to turbidity?

Only solid, legitimate, careful scientific analysis may provide the answer.

One thing is certain; before we leap into decisions that are ill-advised, that are based on anything but sound science, and that "feel" good, we need to step back and size up the situation. This sizing-up procedure can be correctly accomplished only through the use of scientific methods.

Don't we already have too many dysfunctional managers making too many dysfunctional decisions that result in harebrained, dysfunctional analysis—and results?

Obviously, there is no question that we need to stop the pollution of Chesapeake Bay.

However, shouldn't we replace the timeworn and frustrating position that "we must start somewhere" with good common sense and legitimate science?

The bottom line: We shouldn't do anything to our environment until science supports the investment. Shouldn't we do it right?

Frank R. Spellman

THE BOTTOM LINE ON MIGRATION AND INFECTIOUS DISEASE

The potential impact of population mobility on health and on use of health services of migrant host nations is increasing in its importance and cost—and definitely being felt in the United States at the present time (2022) with its open-border policy. It is important to keep in mind that when migrants migrate due to climate change, the fact that they have been exposed to climate change conditions could mean that they have contracted disease, and the disease accompanies them as they enter the migrant receiving area. Moreover, the mobility of migrants and their accompanying disease(s) is not limited to people; animals, insects, birds, and others are mobile too and may bring with them undesirable passengers. There is little doubt that climate change will have significant impacts on both human and animal migration and population health, including infectious disease.

REFERENCES

CDC. (2021). Our risk for infectious diseases is increasing because of climate change. Accessed 02/03/2022 @ www.cdc.gov/ncezid.

CDC. (2022). Animals and rabies. Accessed 02/03/2022 @ www.cdc.gov/rabies/animals/index.html.

Gurarie, E., Hebblewhite, M., Joly, K., Kelly, A.P., Adamczewski, J., Davidson, S.C., Davison, T., Gunn, A., Suitor, M.J., Fagan, W.F., and Boelman, N. (2019). Tactical departures and strategic arrivals: Divergent effects of climate and weather on caribou spring migrations. *Ecosphere* 10(12):e02971.

Spellman, F.R. (2000). *Microbiology for Water and Wastewater Operators*. Boca Raton, FL: CRC Press.

Vasconcelos, V.M., and Pereira, E. (April 2001). Cyanobacteria diversity and toxicity in a wastewater treatment plant (Portugal). *Water Research* 35(5):1354–1357.

6 When the Seas Rise

MAJOR PHYSICAL EFFECTS OF SEA LEVEL RISE

With increased global temperatures, global sea level rise will occur at a rate unprecedented in human history (Edgerton, 1991). Changes in temperature and sea level will be accompanied by changes in salinity levels. For example, a coastal freshwater aquifer is influenced by two factors: pumping and mean sea level. In pumping, if withdrawals exceed recharge, the water table is drawn down and saltwater penetrates inland. With mean sea level, the problem occurs if sea level rises and the coastline moves inland, reducing aquifer area. Additional problems brought about by changes in temperature and sea level are seen in tidal flooding, oceanic currents, biological processes of marine creatures, runoff and landmass erosion patterns, and saltwater intrusion.

You can easily see one of the most important direct physical effects of sea level rise on a coastal beach system. At current rates of sea level rise of 1 to 2 mm (about 0.08 in)/year, significant coastal erosion is already produced. Two major factors contribute to beach erosion. First, deeper coastal waters enhance wave generation, thus increasing their potential for overtopping barrier islands. Second, shorelines and beaches will attempt to establish new equilibrium positions according to what is known as the *Bruun rule*; these adjustments will include a recession of shoreline and a decrease in shore slope (Bruun, 1986).

MAJOR DIRECT HUMAN EFFECTS OF SEA LEVEL RISE

Along with the physical effects of sea level rise, in one way or another, directly or indirectly, accompanying effects have a direct human side, especially concerning human settlements and the infrastructure that accompanies them: highways, airports, waterways, water supply and wastewater treatment facilities, landfills, hazardous waste storage areas, bridges, and associated maintenance systems. Sea level rise could also cause intrusion of saltwater into groundwater suppliers (Edgerton, 1991).

To point out that this infrastructure will be placed under tremendous strain by a rising sea level coupled with other climatic change is to understate the possible consequences. Indeed, the impact on infrastructure is only part of the direct human impact. For example, there is widespread agreement among scientists that any meaningful change in world climate resulting from warming or cooling will (1) disrupt world food production for many years, (2) lead to a sharp increase in food prices, and (3) cause considerable economic damage.

GLOBAL WARMING AND SEA LEVEL RISE

In the past few decades, human activities (burning fossil fuels, leveling forests, and producing synthetic chemicals such as CFCs) have released into the atmosphere huge

DOI: 10.1201/9781003295211-6

quantities of carbon dioxide and other greenhouse gases. These gases are warming the earth at an unprecedented rate. If current trends continue, they are expected to raise earth's average surface temperature by at least 1.5°C to 4.5°C (or more) in the next century—with warming at the poles two to three times as high as warming at the middle latitudes (Spellman and Whiting, 2006).

If we assume global warming is inevitable and/or is already underway, what, then, must we do? Obviously, we cannot jump off the planet and head for greener pastures. We live on Earth and are stuck here (we have no effective method or technology to allow us to leave or a convenient place to go if we tried). If this is the case, understanding the dynamics of change that are evolving around us and taking whatever prudent actions we can to mitigate the situation makes good sense.

We must also take this attitude and approach with the effect global warming is having on the rise in sea level. This rise is already underway, and with it will come increased storm damage, pollution, and subsidence of coastal lands.

"Rise in sea level is already underway?" Absolutely. Consider the following information taken from USEPA's (2017) report, *The Probability of Sea Level Rise*.

1. Global warming is most likely to raise sea levels 15 cm (about 5.91 in) by the year 2050 and 34 cm (about 1.12 ft) by the year 2100. There is also a 10 percent chance that climate change will contribute 30 cm (about 11.81 in) by 2050 and 65 cm (about 2.13 ft) by 2100. These estimates do not include sea level rise caused by factors other than greenhouse warming.

2. There is a 1 percent chance that global warming will raise sea level 1 meter in the next 100 years and 4 meters in the next 200 years. By the year 2200, there is also a 10 percent chance of a 2-meter contribution. Such a large rise in sea level could occur either if the Antarctic Ocean temperature warms 5°C and Antarctic ice streams respond more rapidly than most glaciologists expect, or if Greenland temperatures warm by more than 10°C. Neither of these scenarios is likely.

3. By the year 2100, climate change is likely to increase the rate of sea level rise by 4.1 mm (about 0.16 in)/yr. There is also a one in ten chance that the contribution will be greater than 10 mm (about 0.39 in)/yr, as well as a one in ten chance that it will be less than 1 mm (about 0.04 in)/yr.

4. Stabilizing global emissions in the year 2050 would be likely to reduce the rate of sea level rise by 28 percent by the year 2100 compared with what it would be otherwise. These calculations assume that we are uncertain about the future trajectory of greenhouse gas emissions.

5. Stabilizing emissions by the year 2025 could cut the rate of sea level rise in half. If a high global rate of emissions growth occurs in the next century, sea level is likely to rise 6.2 mm (about 0.24 in)/yr by 2100; freezing emissions in 2025 would prevent the rate from exceeding 3.2 mm (about 0.13 in)/yr. If less emissions growth were expected, freezing emissions in 2025 would cut the eventual rate of sea level rise by one-third.

6. Along most coasts, factors other than anthropogenic climate change will cause the sea to rise more than the rise resulting from climate change alone. These factors include compaction and subsidence of land, groundwater depletion,

and natural climate variations. If these factors do not change, global sea level is likely to rise 45 cm (about 1.48 ft) by the year 2100, with a 1 percent chance of a 112 cm (about 3.67 ft) rise. Along the coast of New York, which typifies the United States, sea level is likely to rise 26 cm (about 10.24 in) by 2050 and 55 cm (about 1.8 ft) by 2100. There is also a 1 percent chance of a 55 cm (about 1.8 ft) rise by 2050 and a 120 cm (about 3.94 ft) rise by 2100.

Along with the EPA's findings reported previously, additional lines of evidence corroborate that global mean sea level has been rising during at least the last 100 years; this evidence is apparent in tide gauge records, erosion of 70 percent of the world's sandy coasts and 90 percent of America's sandy beaches, and the melting and retreat of mountain glaciers. Edgerton (1991) points out that the correspondence between the two curves of rising global temperatures and rising sea levels during the last century appears to be more than coincidental.

Major uncertainties are present in estimates of future sea level rise. The problem is further complicated by our lack of understanding of the mechanisms contributing to recent rises in sea level. In addition, different outlooks for climatic warming dramatically affect estimates. In all this uncertainty, one thing is sure. Estimates of sea level rise will undergo continual revision and refinement as time passes and more data are collected.

General Question 6.1—Just how much of a rise in sea level are we talking about? According to USEPA (2017), "if the experts on whom we relied fairly represent the breadth of scientific opinion, the odds are fifty-fifty that greenhouse gases will raise sea level at least 15 cm by the year 2050, 25 cm by 2100, and 80 cm by 2200" (p. 123).

RESPONSE SCENARIO TO OTHER GENERAL QUESTIONS 6.1— GLOBAL CLIMATE CHANGE

Is global warming a hoax? Is Earth's climate changing? Are warmer times or colder times on the way? Is the greenhouse effect going to affect our climate, and if so, do we need to worry about it? Will the tides rise and flood New York? Does the ozone hole portend disaster right around the corner?

These and many other questions related to climate change have come to the attention of us all. We are inundated by a constant barrage of newspaper headlines, magazine articles, and television news reports on these topics. Recently, we've seen report after report on El Niño and its devastation of the west coast of the United States (and Peru and Ecuador)—and its reduction of the number, magnitude, and devastation of hurricanes that annually blast the east coast of the United States.

To illustrate just how constant the barrage of newspaper headlines has been, we have selected and listed a sample of the climate change news and global warming headlines published in many locations throughout the globe in the month of April 2008 (Carbonify.com, 2009).

April 1, 2008—GLOBAL WARMING AWARENESS AND APATHY
April 2, 2008—OCEANS UNDER STRESS FROM GLOBAL WARMING
April 7, 2008—AUSTRALIAN DROUGHT AFFECTED AREAS GROW

April 13, 2008—FOSSIL FUEL CARBON EMISSIONS OVER 8 GIGATONS
April 22, 2008—UK MIGRATING BIRDS NUMBERS DROP
April 28, 2008—MARCH WARMEST ON RECORD GLOBALLY

Scientists have been warning us of the catastrophic harm that can be done to the world by atmospheric warming. One view states that the effect could bring record droughts, record heat waves, record smog levels, and an increasing number of forest fires.

Another caution put forward warns that the increasing atmospheric heat could melt the world's icecaps and glaciers, causing ocean levels to rise to the point where some low-lying island countries would disappear, while the coastlines of other nations would be drastically altered for ages—or perhaps for all time.

What's going on? We hear plenty of theories put forward by doomsayers, but are they correct? If they are correct, what does it all mean? Does anyone really know the answers? Should we be concerned? Should we invest in waterfront property in Antarctica? Should we panic?

No. While no one really knows the answers—"we don't know what we don't know syndrome"—and while we should be concerned, no real cause for panic exists.

Should we take some type of decisive action—should we produce quick answers and put together a plan to fix these problems? What really needs to be done? What can we do? Is there anything we can do?

The key question to answer here is "What really needs to be done?" We can study the facts, the issues, the possible consequences—but the key to successfully combating these issues is to stop and seriously evaluate the problems. We need to let scientific fact, common sense, and cool headedness prevail. Shooting from the hip is not called for, makes little sense—and could have titanic consequences for us all.

The other question that has merit here is, "Will we take the correct actions before it is too late?" The key words here are: "correct actions." Eventually, we may have to take some action (beyond hiding in a cave somewhere). But we do not yet know what those actions could be or should be.

From our perspective, one thing is certain: in our college-level environmental health courses, we address, eventually, global warming and/or global climate change. Through time and experience we have learned (yes, teachers learn, too) that whether we call it global warming, global climate change (humankind-induced global warming, under a broader label), or an inconvenient truth, the topic is a conundrum (a riddle, the answer of which is a pun). As such, before diving into the many emotionally charged, heated class discussions about this "hot" topic (pun intended).

Consider this: any damage we do to our atmosphere affects the other three environmental mediums: water and soil—and biota (us—all living things). Thus, the endangered atmosphere (if it is endangered) is a major concern (a life-and-death concern) to all of us.

WHEN THE ICE MELTS

During this current episode of climate change and the warming that accompanies it, the ice is melting.

The ice is melting?

Yes.

Where is the ice melting?

Okay, those questions are easy to answer. First, let's begin with the February 3, 2022, CNN report about Mt. Everest and the highest glacier on the world's tallest mountain losing decades worth of ice every year because of global climate change. Dewan and Gainor (2022) reported that ice that took around 2,000 years to form on the South Col Glacier has melted in around 25 years, which means it has thinned out around 80 times faster than it formed.

Okay, what else, you ask?

One only need study the Glacier National Park receding glaciers and the Alaska Matanuska glacier to quickly answer this question. In 2011, my partner JoAnn Garnett-Chapman and I surveyed the Matanuska glacier and were shocked at what we saw. As shown in Figures 6.1 through 6.3, the glacier is not only melting but is moving into the sea at a quicker pace than we measured years earlier. That part of our survey of glacier, which is 27 miles in length, did not surprise us, but what did is that it appeared to have shrunk in width somewhat. Although the glacier is easily accessible via Glacier Road, and we had no problem hiking its length (it took a few days) and breadth, we were not able to judge the shrinkage in a scientific manner but only by eyeballing it as we walked along. Figure 6.4 shows what remains of the

FIGURE 6.1   Photo showing the Matanuska Glacier, Alaska.

Source: Photo by F.R. Spellman and JoAnn Chapman (2011).

FIGURE 6.2   Photo showing the Matanuska Glacier, Alaska, entering the sea.

Source: Photo by F.R. Spellman and JoAnn Chapman (2011).

FIGURE 6.3   Photo showing the Matanuska Glacier, Alaska, and melt area.

Source: Photo by F.R. Spellman and JoAnn Chapman (2010).

FIGURE 6.4   Photo showing the remnants of Elizabeth Lake Glacier in 2009 in Glacier National Park, Montana.

Source: Photo by F. Spellman.

FIGURE 6.5   Image showing Thwaites Glacier and ice shelf.

Source: NOAA (2022). Accessed 02/05/2022 @ https://earthobservatory.nasa.gov/images/146247/thwaites-glacier-transformed.

Glacier National Park, Montana, Elizabeth Lake glacier as of 2009. Anyway, the point is, we both knew that thousands of gallons of ancient water was now pouring into the sea and mean sea level rise was occurring worldwide, and relative sea level rise too—relative sea level rise is when the sea rises and the land subsides, like is presently occurring in Hampton Roads, a.k.a. Tidewater, Virginia.

There is more. When the glaciers melt, recede, fall into the seas or lakes or wherever, we are talking about massive amounts of water being dumped and increased water levels, baring land areas. For those of us concerned with global climate change and migrations, this is important. Simply, change brings about change, and when change is detrimental to our very existence, it can be viewed as important to those who think about such things.

THE DOOMSDAY GLACIER

The Doomsday Glacier (a.k.a. Thwaites Glacier) is a vast Antarctic glacier flowing into Pine Island Bay, part of the Amundsen Sea on the Walgreen Coast of Marie Byrd Land, Antarctica. Its measured surface speed exceeds 1.2 miles (2 kilometers) per year near its grounding line. This glacier is closely watched for its potential to raise sea levels dramatically (Gertner, 2018). It has been described as part of the weak underbelly of the West Antarctic Ice Sheet because of its apparent vulnerability to significant retreat. Substantial changes in this glacier and its surface lowering and grounding line retreat indicate that this is a glacier not to be ignored. We are talking about a glacier, a frozen water hunk the size of Florida—meaning it is an instant sea level rise worry—remember, increased sea levels in coastal settlements can push migrants to seek higher ground. Figure 6.5 shows the Thwaites Glacier. When Thwaites enters the water, it follows a similar blueprint of cave in. At first, water nibbles at the ice shelf from below, causing it to destabilize and to become slender and thin. And instead of sitting securely on the seafloor, it begins to float, like a beached boat lifted off the sand. This is not good because when it floats, it is exposed to more water, and the weakening and thinning not only continue but accelerate. The glacier (the shelf) is now too fragile to support its own weight and starts calving off into the sea in enormous chunks. More water added to the seas, more problems for coastal shorelines. The cycle continues over and over again as more ice flows down from the glacier's interior, replenishing what is lost, meaning increased water levels may be an icy ticket to ride to the promised land(s).

POLAR BEARS AND CLIMATE CHANGE

Again, it is important to remember that global climate affects more than just humans; other living organisms are affected also. Besides the organisms described earlier and their response to climate change, one of the animals that has received a lot of attention with regard to global warming and ice melt is the polar bear (see Figure 6.6). The problem is that as a major reduction in sea ice habitat occurs, the polar bears' supply of food, which is seals (their main prey), diminishes. With the loss of their food

FIGURE 6.6 Polar bears (*Ursus martimus*).

Source: U.S. Fish and Wildlife Service. Public domain library photo by Gary Kramer. Accessed 02/07/2022 @ https://digitalmedia.fws.gov/digital/collection/natdiglib/id/14121/rec/1.

supply, it is likely that the overall population of polar bears and subpopulations could ultimately lead to further reduction in numbers. It is likely polar bears will migrate to more opportunistic locations or their numbers will decline.

REFERENCES

Bruun, P. (1986). Worldwide impacts of sea level rise on shorelines. In *Effects of Changes in Stratospheric Ozone and Global Climate*, vol. 4. New York: UNEP/EPA, pp. 99–128.

Carbonify.com. (2009). Global warming–a hoax? Accessed 11/07/2009 @ www.carbonify. Com/articles/global-warming-hoax.htm.

Dewan, A., and Gainor, D. (2022). Ice melt: Mt. Everest. Accessed 02/05/2022 @ www.cnn. com/2022/02/03/asia/mt-everest.

Edgerton, L. (1991). *The Rising Tide: Global Warming and World Sea Levels*. Washington, DC: Island Press.

Gertner, J. (2018). The race to understand Antarctica's most terrifying glacier. Accessed 02/05/2022 @ www.wired.com/story/antarctica-thwaites-glacier-breaking-point.

Spellman, F.R., and Whiting, N. (2006). *Environmental Science and Technology: Concepts and Applications*. Boca Raton, FL: CRC Press.

USEPA. (2017). *The Probability of Sea Level Rise*. Washington, DC: Environmental Protection Agency.

7 When the Forest Turns to Ash

WILDFIRES!

Wow, think about it: wildfires have been the subject of legends; the source of sustenance of many of America's forests, shrub lands, grasslands, and other wildlands; and a central theme in forest management in the United States during the 20th century (Agee, 1998). Fire is important, especially in many ecosystems where it plays a vital role (Hutto, 2008), in part by promoting a montage of vegetation and by stimulating the establishment and growth of particular trees and other plants (Brown, 2000). Truly, many wildland species, such as lodgepole pine found in the West, require fire to regenerate (Brown and Smith, 2000). However, wildfire can also cause economic and ecological losses; can pose threats to people, property, and communities; and can result in displacement (a.k.a. migration).

One of the current occurrences that helps to make wildfires so damaging is the increasing movement of people to live in and around forests, grasslands, shrub lands, and other natural areas—the getting back to nature movement, and away from the crime, crud, collapse, and homelessness of the city environment. These places of settlement are commonly referred to as the wildland–urban interface, or WUI— where fire-related challenges are on the increase (Hammer et al., 2009; NASF, 2009). Exacerbating the wildland fire problem is that the number of wildfires exceeding 50,000 acres has increased over the past 30 years, with most of the change occurring over the past 10 years (NASF, 2009). Many of these large wildfires are more intense and damaging than in the past (Hardy et al., 2001).

DID YOU KNOW?

The range of ecological processes and conditions that characterized various ecosystems in the United States prior to European settlement, referred to as "historical range of variability" (HRV), has been a subject of much research (Keane et al., 2009). HRV is used by managers and scientists as a reference point to assess current conditions.

As more people choose to work or live in the WUI, fire management becomes more complex and dangerous for residents. Moreover, there is also a cost factor: the costs to reduce the risk, fight wildfires, and protect homes and human lives (and animals) have risen sharply in recent decades (Abt et al., 2009; NASF, 2009). Climate

DOI: 10.1201/9781003295211-7

change, insect pests, and diseases, among other influences, are also contributing to vast changes in wildland vegetation that in many areas result in landscapes that are drier, less resilient, and more likely to burn once ignited (Keane et al., 2008a, 2008b)—a significant factor in displacement (a.k.a. forced migration).

As of 2005, some 32+ percent of U.S. housing units and one-tenth of all land with housing are situated in the wildland-urban interface, and WUI growth is expected to continue (Hammer et al., 2009).

CLIMATE CHANGE INDICATOR AND MIGRATION DRIVER: WILDFIRES

Following 2020's extreme fire season, high-elevation forests in the central Rocky Mountains are now burning more than at any point in recorded history—more so than in the last 2,000 years—according to a new University of Montana study (2021). The research team used charcoal core samples found in lake sediments to piece together the fire history of forests across the Rocky Mountains. The forests surrounding National Parks are not immune to wildfire and actually are more prone to damage from wildfires than many other locations (see Figure 7.1).

FIGURE 7.1   Keauhou fire in Hawai'i Volcanoes National Park in 2018.

Source: US National Park Service. Public Domain. USGS Wildland Fire Science. Accessed 05/08/2022 @ www.usgs.gov/media/images/keauhou-fire-hawaii-volcanoes-naonal-park-2018.

Let's look at a few relatively current key points about wildfires (USEPA, 2021):

- Since 1983, the National Interagency Fire Center has documented an average of approximately 70,000 wildfires per year. Based on data compiled from the Forest Service, the actual total may be even higher for the data that can now be compared. Note that the data do not show an obvious trend during this time.
- Since the 1980s, the extent of the area burned by wildfires each year appears to have increased. Note that in the 2004 to 2014 timeframe, the largest acreage burned, with a peak year in 2015. This coincides with many of the warmest years on record nationwide (USEPA, 2021).
- In the late 1990s there was a period of transition in certain climate cycles that tend to shift every few decades (Fann et al., 2018). This shift—combined with other ongoing changes in temperature, drought, and snowmelt—may have contributed to warmer, drier conditions that have fueled wildfires in parts of the western United States (Kitzberger et al., 2007; Westerling, 2016).
- The proportion of burned land from 1984 to 2018 that suffered severe damage ranged from 5 to 13 percent.
- The actual land area burned by wildfires varies by state. Fires burn more land in the western United States than in the East, and parts of the West and Southwest show the largest increase in burned acreage between the first half of the period of record (1984–2001) and the second half (2002–2018).
- The peak of the U.S. wildfire season is occurring earlier. In 1984–2000, burned area peaked in August. More recently, it has peaked in July. An average of 1.6 million acres burned in July of each year from 2001 to 2017 (2021).

Collectively, forests, shrublands, and grassland cover more than half of the land area in the United States (MRLC, 2019). While it is true that wildfires occur naturally and play a long-term role in the health of these ecosystems, changing wildfire patterns threaten to upset the current situation. Several studies have found that climate change has already led to an increase in wildfire season length, wildfire frequency, and burned area (Westerling, 2016; USGCRP, 2018). Because of factors including warmer springs, longer summer dry seasons, and drier soils and vegetation, the wildfire season has lengthened in many areas (USGCRP, 2018). Also, climate change threatens to increase the frequency, extent, and severity of fires through increased temperatures and drought (USGCRP, 2018). Earlier spring melting and reduced snowpack, one of the climate change indicators that USEPA's (2021) report details, result in decreased water availability during scorching summer conditions, which in turn contributes to an increased wildfire risk, allowing fires to start more easily and burn hotter. The specific points USEPA's (2021) snowpack report includes:

- From 1955 to 2020, April snowpack declined at 86 percent of the sites measured (western United States). The average change across all sites amounts to about a 19 percent decline.
- Large and consistent decreases in April snowpack have been observed throughout the western United States (USGCRP, 2018). Decreases have

been especially prominent in Washington, Oregon, northern California, and the northern Rockies.

- Although some stations have experienced increases in April snowpacks, all twelve states included in this indicator experienced a decrease in snowpack on average from 1955 to 2020. In the Northwest (Idaho, Oregon, Washington), all but four stations saw decreases in snowpack over the period of record.
- A shift toward earlier peak snowpack has been experienced in about 81 percent of the sites. This earlier trend is especially pronounced in southwestern states like Colorado, New Mexico, and Utah. Across all stations, peak snowpack has shifted earlier by an average of nearly eight days since 1982, based on the long-term average rate of change.

The trend of longer wildfire seasons and larger wildfire size are predicted to continue as more frequent and longer droughts occur (USGCRP, 2018). In addition to climate change, other factors—land use, large-scale insect infestation, fuel availability (including invasive species such as highly flammable cheatgrass; see Figure 7.2), and management practices, including fire suppression—play a key role in wildfire frequency and intensity. A combination of these factors (or those occurring separately) can occur, but they vary greatly by region and over time, as do precipitation, wind, temperature, vegetation types, and landscape conditions. Accordingly, understanding changes in fire characteristics requires long-term records, a regional perspective, and consideration of many factors (Stein et al., 2013).

FIGURE 7.2   Cheatgrass.

Source: USDA (2012). Public domain photo.

DID YOU KNOW?

Cheatgrass and other invasive annual grasses, such as medusa head and vente-nata, are taking over America's sagebrush rangelands, increasing wildfire size and frequency, reducing forage productivity, and threatening wildlife habitat and rural economies. Lack of bold and coordinated action is making our working lands less productive for each passing generation (USDA, 2012).

Wildfires have the potential to harm property, livelihoods, and human and wildlife health. Fire related threats are increasing, especially as more people live in and around forests, grasslands, and other natural areas (National Association of State Foresters, 2009). As reported by the National Oceanic and Atmospheric Administration, between 1980 and 2020 the United States had eighteen wildfire events that caused more than $1 billion in damage; fifteen of those have occurred since 2000 (NOAA, 2021). Over the past few decades, the United States has routinely spent more than $1 billion per year to fight wildfires, including $1.6 billion in 2019 (NIFC, 2020), with no decrease in sight. Unfortunately, these events have resulted in the deaths of more than 1,000 firefighters since 1910 (NIFC, 2019). Even in communities far downwind, wildfire smoke has been linked to poor air quality that can lead to significant health effects and costs to society (emergency department visits, hospital admissions, and deaths, often due to respiratory ailments) (Johnston et al., 2012; Fann et al., 2016; Youssouf et al., 2014; Jones and Berrens, 2017). Keep in mind that when it is stated that wildfires can lead to significant health effects, we are not to forget the animal and wildlife community. The bottom line here is that wildfires can and do lead to displacement, and, except for wildlife that can handle and survive and thrive in wildfire situations, all else, humans included, may need to get out of Dodge and seek the promised lands—wherever they might be.

Beyond the human, societal, and wildlife impacts, wildfires also affect the Earth's climate. So what we are saying here is that climate change leads to increased wildfire events and the wildfires lead to climate change—or increased climate change. Note that forests in particular store copious amounts of carbon. When they burn, they immediately release carbon dioxide into the atmosphere, which in turn contributes to climate change. Keep in mind that that it is not only the burning that releases carbon dioxide to the atmosphere, but carbon is also released as the remnants decompose.

Truth be told, reducing the loss of lives, property, infrastructure, and natural resources all depends on long-term community action (NFPA, 2006). Land use decisions, building codes and standards, and other planning and landscaping choices all influence a community's vulnerability to damage from wildfire (Blonski et al., 2010). Experience has demonstrated that communities can reduce the risk of damage by becoming knowledgeable about and engaged in actions to plan and protect their homes and neighborhoods from wildfires. Such fire-adapted communities will be better prepared to safely accept wildfires as part of their surrounding landscape (Leschak, 2010; NASF, 2009; NFPA, 2006).

CLIMATE CHANGE–DRIVEN MIGRATION—
COMPLEX AND CONTROVERSIAL

If you accept the premise that the increase in wildland fires is climate change driven, then you might ask why is there, at the present time, so much controversy, argument, and dispute related to what you might think is the obvious? The obvious? Yes, it involves controlled burning (a.k.a. prescribed burning; see Figure 7.3) in forests to reduce fuel load. While it is true that many people experience and interpret wildfire in terms of the damage caused to human lives, structures, and communities, it is also important to recognize the ecological role fire plays across landscapes. Wildfire is a fundamental ecological component for more than 90 percent of wildlands across the conterminous United States (USDA Forest Service, 2012). Fire-adaptable (i.e., having an ability to adjust to the intensity and quality of disturbance—fire) ecosystems and species are found in every corner of the United States, from the Pacific Northwest to the Rocky Mountain West, to the Southwest's chaparral, the Midwest's tall prairies, the New Jersey pine barrens, and the South's longleaf pine forests. On the other hand, many conservationists argue that controlled burns are not that effective in slowing down a bushfire and that they damage natural habitat. The conservationists state that wildland fires are best orchestrated as Mother Nature designs and dictates.

FIGURE 7.3   Great Plains controlled burn.

Source: National Park Service (2018). Public domain photo.

We will leave these opinions to others. The point being made herein is that wildland fires displace humans and wildlife and basically turn them into migrants looking for greener pastures, so to speak.

REFERENCES

Abt, K.L., Prestemon, J.P., and Gobert, K.M. (2009). Wildfire suppression cost forecast for the US Forest Service. *Journal of Forestry* 107(4):173–178.

Agee, J.K. (1998). The landscape ecology of western forest fire regimes. *Northwest Science* 72(special issue):24–34.

Blonski, K., Miller, C., and Rice, C. (2010). *Managing Fire in the Urban Wildland Interface*. Point Arena, CA: Solano Press Books.

Brown, J.K. (2000). Introduction and fire regimes. In: Brown, J.K. and Smith, K.K., editors. *Wildland Fire in Ecosystems: Effects of Fire on Flora*. Gen Tech MRS-GTR-42, Vol. 2. Ogden, UT: U.S. Department of Agriculture, Forest Service, Rocky Mountain Research Region, pp. 1–7.

Brown, J.K., and Smith, J.K. (2000). *Wildland Fire in Ecosystems: Effects of Fire on Flora*. Ogden, UT: U.S. Department of Agriculture, Forest Service, Rocky Mountain Research Region.

Fann, N., Alman, B., Broome, R.A., Morgan, G.G., Johnston, F.H., Pouliot, G., and Rappold, A.G. (2018). The health impacts and economic value of wildland fire episodes in the U.S.: 2008–2012. *Sci. Total Environ.* 610–611:802–809.

Fann, N., Brenna, T., Dolwick, P., Gamble, J.L., Ilacqua, V., Kolb, L., Nolte, C.G., Spero, T.L., and Ziska, L. (2016). Chapter 3: Air quality impacts. In: *The Impacts of Climate Change on Human Health in the United States: A Scientific Assessment*. U.S. Global Change Research Program. Accessed 02/14/2022 @ https://health2016.globalchange.gov.

Hammer, R.B., Stewart, S.I., and Radeloff, V.C. (2009). Demographic trends, the wildland-urban interface, and wildfire management. *Society & Natural Resources* 22(8):777–782.

Hardy, C.C., Schmidt, K.M., Menakis, J.P., and Sampson, R.N. (2001). Spatial data for national fire planning and fuel management. *International Journal of Wildland Fire* 10(4):353–372.

Hutto, R.L. (2008). The ecological importance of severe wildfires: Some like it hot. *Ecological Applications* 18(8):1827–1834.

Johnston, F.H., Henderson, S.B., Chen, V., Randerson, J.T., Marlier, M., DeFries, R.S., Kinney, P., Bowman, D., and Brauer, M. (2012). Estimated global mortality attributable to smoke from landscaper fires. *Environ. Health Persp.* 120(5):695–701.

Jones, B.A., and Berrens, R.P. (2017). Application of an original wildfire smoke health cost benefits transfer protocol to the western U.S. 2005–2015. *Environ. Manage.* 60:809–822.

Keane, R.E., Agee, J.K., Fule, P., Keeley, J.E., Key, C., Kitchen, R., and Schulte, L.A. (2008a). Ecological effects of large fires on US landscapes: Benefit or catastrophe? *International Journal of Wildland Fire* 17:169–712.

Keane, R.E., Hessberg, P.R., Landres, P.B., and Swanson, F.J. (2009). The use of historical range and variability (HRV) in landscape management. *Forest Ecology and Management* 258:1025–1037.

Keane, R.E., Holsinger, L.M., Parsons, R.A., and Gray, K. (2008b). Climate change effects on historical range and variability of two large landscapes. *Forest Ecology and Management* 254:375–389.

Kitzberger, T., Brown, P.M., Heyerdahl, E.K., Swetnam, T.W., and Veblen, T.T. (2007). Contingent Pacific-Atlantic ocean influence on multicentury wildfire synchrony over western North America. *P. Natl. Acad. Sci. USA* 104(2):543–548.

Leschak, P. (2010). Strong partnerships and the right tools: The pre-wildfire strategy of fire adapted communities. *Disaster Safety Review* 8:4–5.

MRLC (Multi-Resolution Land Characteristics). (2019). National land cover database 2016 statistics for 2016. Accessed 02/14/2022 @ www.mrlc.gov/data/statistics/national-land-cover-database-2016-nicd2016-statistics.

National Association of State Foresters. (2009). Quadrennial fire review. Accessed 02/14/2022 @ www.forestandranglands.gov/documents/strategy/foundational/qfr2009final.pdf.

NIFC (National Interagency Fire Center). (2019). Wildland fire fatalities by year (1910–2017). Accessed 02/14/2014 @ www.nifc.gov/fire-information/statistics/suppression-costs.

NIFC (National Interagency Fire Center). (2020). Historical wildland fire information: Federal firefighting costs: Suppression only (1985–2019). Accessed 02/14/2022 @ www.nifc.gov/fire-information/statistics/suppression-costs.

NFPA (National Fire Protection Association). (2006). Firewise: Community solutions to a national problem. National Wildland/Urban Interface fire program publication FWC-008–06-BK. Quincy, MA, 39 p.

NOAA (National Oceanic and Atmospheric Administration). (2021). Billion-dollar weather and climate disasters. Accessed 02/14/22 @ www.ncdc.noaa.gov/billions.

Stein, S.M., Menakis, J., Carr, M.A., Comas, S.J., Stewart, S.I., Cleveland, H., Bramwell, L., and Radeloff, V.C. (2013). Wildfire, wildlands, and people: Understanding and preparing for wildfire in the wildland-urban interface. Gen. Tech. Rep. RMRS-GTR-299. Fort Collins, CO: U.S. Department of Agriculture. Accessed @ www.fs.fed.us/openspace/fote/wildfire-report.html.

University of Montana. (2021). Rocky mountain forest burning. Accessed 12/21/21/ @ https://www.umt.edu/news/2021/o6/061421fire.php.

USDA (United States Department of Agriculture, Forest Service). (2012). *Historical Natural Fire Regimes Map*. Unpublished analysis of LANDFIRE:1.05 fire regime groups layer. On file G. Dillon, U.S. Forest Service, Rocky Mountain Research Station, Fire Modeling Institute, Missoula Fire Sciences Laboratory, 5775 W. Highway 10, Missoula, MT, 59808–9361.

USEPA. (2021). Climate change indicators: Wildfires. Accessed 02/10/2022 @ https://epa.gov/sites/default/files/2021-04/wildfires.

USGCRP (U.S. Global Change Research Program). (2018). *Impacts, Risks, and Adaptation in the United States: Fourth National Climate Assessment*, Vol. 2. Reidmiller, D.R., Avery, C.W., Easterling, D.R., Kunkel, K.E., Lewis, K.L.M., Maycock, T.K., and Stewart, B.C., editors. Accessed 02/10/2022 @ https://nca2018.globalchange.gov.

Westerling, A.L. (2016). Increasing western U.S. forest wildfire activity: Sensitivity to changes in the timing of spring. *Phil. Trans. R. Soc. B* 371:20150178.

Youssouf, H., Liousse, C., Roblou, L., Assamoi, E.-M., Salonen, R.O., Maesano, C., Banerjee, S., and Annesi-Maesano. I. (2014). Non-accidental health impacts of wildfire smoke. *Int. J. Environ. Res. Public Health* 11:11772–11804.

SUGGESTED READINGS

USGCRP (U.S. Global Change Research Program). (2017). *Climate Science Special Report: Fourth National Climate Assessment*, Vol. 1. Wuebbles, D.J., Fahey, D.W., Hibbard, K.A., Dokken, D.J., Stewart, B.C., and Maycock, T.K., editors. Accessed 02/14/2022 @ https://science2017.globalchange.gov.

8 When the Mountains Lift

HELL ON EARTH

It was horrendous. It was a lateral blast . . . it was a horizontally directed explosion of incredible magnitude. It caused this expanding cloud of ash, rocks, and gases to move out across the countryside to the north at speeds of several hundreds of miles an hour. The directed blast was really the most destructive event that occurred on the morning of May 18. It completely destroyed an area of 230 square miles in a matter of somewhere between five and nine minutes. It essentially killed every living thing within an area of 230 square miles (more than 7,000 mammals, including elk and deer, were killed along with 57 humans—the kill zone, directly on a north track, was 14 miles in length). And it destroyed hundreds of acres of virgin forest, an incredibly catastrophic event—literally a hell on Earth[1] (see Figure 8.1).

FIGURE 8.1 Eruption of Mt St Helens on May 18, 1980, shown from southwest aerial view.

Source: USGS Public Domain Photo. Accessed 0/05/2022 @ www.usgs.gov/media/images/plinian-column-may-18-1980-eruption-mount-st-helens-aeri.

DOI: 10.1201/9781003295211-8

I have studied this location and the catastrophic event that took place here in May of 1980 for more than 40 years. Almost 42 years ago, after two months of earthquakes and small explosions, Mount St. Helens cataclysmically erupted. The high-speed blast leveled millions of trees and ripped soil from bedrock. The eruption fed a towering plume of ash for more than nine hours, and winds carried the ash hundreds of miles away. Volcanic mudflows (lahars) carried large boulders and logs, which destroyed forests, bridges, roads, and buildings. This catastrophic event led to 57 human deaths and more than 7,000 large mammal deaths (elk and deer) and caused the worst volcanic disaster in the recorded history of the conterminous United States (USGS, 2020).

In the past 40 years I have hiked the 33-mile Loowit Trail around the base of Mt. St. Helens four times during different years (see Figures 8.2, 8.3, 8.4) and hiked the Truman Trail (named after Harry R. Truman, who died along with his 18 cats in his lodge at Spirit Lake during the May 18, 1980, eruption) dozens of times in the past 40 years (see Figure 8.5). Not an easy hike but enlightening, educational, and very instructive to me. I deliberately studied the north flank of the mountain. This is the area that had my interest because it is on this side where the flank laterally blew almost horizontally during the eruption. Something new? Well, most of us think that

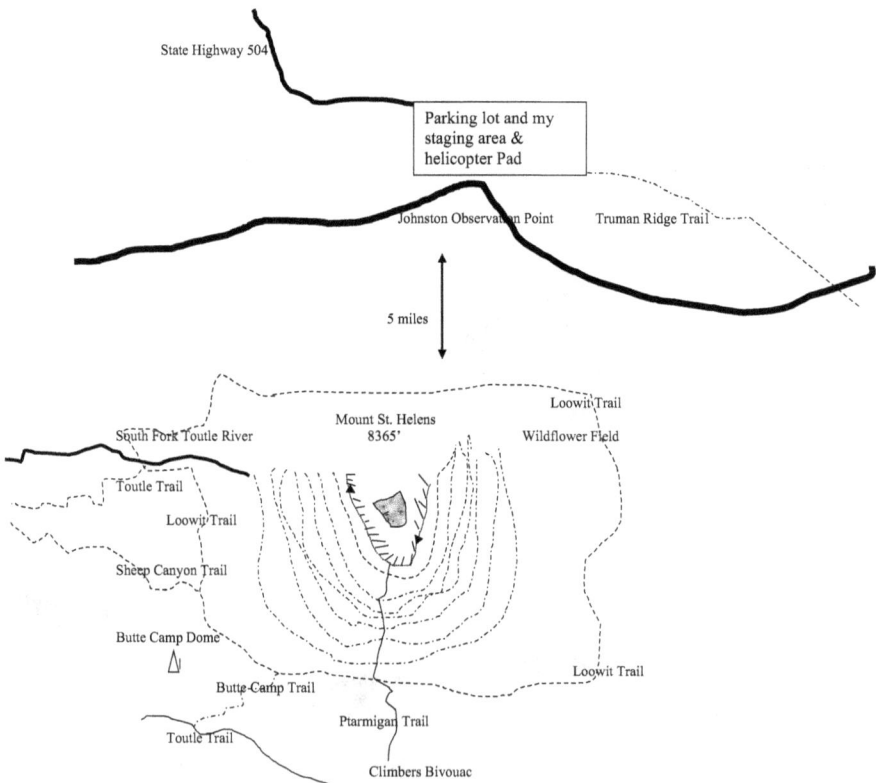

FIGURE 8.2   Rough map of the Loowit Trail and surrounding area.

FIGURE 8.3 Mt St Helens view at northern aspect where the lateral eruption created a 14-mile kill zone directed northward. Loowit Trail follows the base of the mountain.

Source: Photo by F. Spellman (2020).

FIGURE 8.4 The author viewing Mt St Helens from Johnston Observation Point. Taken in 2020 by author who climbed the south face twice and on the second attempt made it to the top.

Source: Photo by F. Spellman (2020).

FIGURE 8.5 Author on Truman Trail with Mt St. Helens in background. Trail leads to over-view of Spirit Lake and the burial grounds of Harry R. Truman and his 18 cats.

Source: Photo by F. Spellman (2020).

when a volcano erupts, it is all about a vertical eruption. But not in this case. This north flank eruption had my interest because it illustrated basically how to interpret the hummocky terrain therein and also near other Cascade volcanoes, such as Mt. Shasta in California—especially the hummocky terrain of the debris avalanche north of Mt. Shasta. What this taught me was that this type of terrain is evidence of a past flank collapse that occurred at Mt Shasta probably around 350,000 years ago and did so without eruption. So what is new is old, and what is old, we are discovering anew.

Since the eruption of Mt. St. Helens, the monitoring of volcanoes has evolved. Today we are able to use scientific instruments to measure earthquakes, deformation, and volcanic gases. The bottom line: we can detect changes in the Earth's surface which may enable us to determine if an eruption is in the cards, so to speak.

The explosive eruption of May 18, 1980, illustrates the importance of developing tools for measuring ground deformation at explosive volcanoes and other unstable areas. Today (2022) we utilize technology advances like photogrammetry, geographic information systems (GISs), and light detection and ranging (LIDAR), all of which enable scientists to make precise measurements and illustrations of changes to Earth's surface, including inflation and deflation at volcanoes. The good news is that technology is dynamic, and advances in telemetry, broad-based seismometer technology, and low-power instrumentation are fueling the dynamism of volcano monitoring equipment capable of collecting and transmitting real-time data remotely with increased precision, efficiency, portability, value, and reduced risk to vulcanologists and others.

AND THE STREAMS WILL TELL

You could not step twice into the same rivers, for other waters are ever flowing on to you.

Heraclitus of Ephesus

You know when and if we ever think about how Earth was formed and how it adjusts and how Mother Nature sculpts the landscape willy-nilly or by her master plan, we may imagine, suppose, presume, or just flat out guess that events such as the Mt St Helens eruption are her major tool, device, and/or mechanism to build up and tear down the global landscape.

Well, there is some truth to this way of thinking, but there is also another side of the coin. It is quite obvious how the landscape changed when Mt St Helens erupted, and Mother Nature used her gigantic hand to re-sculpt the local area (see Figures 8.6 and 8.7) and the follow-up damage or alterations that are still in play at this very time and for some time in the future.

Still in play?

Yes.

What are the elements still in play due to the May 18, 1980, eruption of Mt St Helens?

We will get to the elements, components, constituents, and features still in play in shaping global landscapes momentarily, but for now let's look at one way in which Mother Nature sculpted the global landscape during the 1980 Mt St Helens eruption.

FIGURE 8.6 One of the remnants of the Mt St Helens eruption.

Source: Photo by F. Spellman.

FIGURE 8.7   Inside Mt St Helens' crater hovering above the lava dome in a helicopter.
Source: Photo by F. Spellman (2020).

First, see Figure 8.6, where it shows only one victim of the blast—remember, millions of old growth trees, shrubs, and mammals were erased or reshaped by the massive explosion. Also, Figure 8.7 is a photograph of (2020) Mt St Helens' crater today with its growing lava dome. This is the area where the mountain's perfectly shaped round crater existed before the eruption. This was a white-knuckled flight by helicopter I will never forget.

So let's describe one way that Mother Nature sculpted the Earth in her way, like the artist she is—then we will return to the ongoing Mt St Helens eruption and its ongoing sculpting.

Transformation: Mountain to Sediment[2]

Three examples are presented in this section describing how Earth landforms are reshaped, sculpted, chiseled, and fashioned only as Mother Nature can do via her natural transformational processes. These three natural examples of land transformation are important with regard to migration and displacement of both human and wildlife populations (including plant life) because if the locations change, it may be time to seek those greener pastures elsewhere—or maybe not.

Example 1

Early in the spring in a snow and ice-covered high alpine meadow is the time and place where the water cycle continues. The cycle's main component, water, has been held in reserve—literally frozen—for the long dark winter months, but with longer, warmer spring days, the sun is higher, more direct, and of longer duration, and the frozen masses of water respond to the increased warmth. The melt begins with a single drop, then two, then three, then more until the beginning of a high-gradient rill is formed. As the snow and ice melt, the drops join a chorus, a definition of vigor, that continues unending; they fall from their ice-bound lip to the bare rock and soil terrain below. A chunk of Laramide orogeny period rock (a 4,113-mm boulder) formed 80 to 55 mya (million years ago, during the late Cretaceous to Paleocene), a small part of the thawing lip (parent rock), finally succumbs to gravity and falls with the cascading drops of melt water. The terrain that the snowmelt and the boulder meet, strike, and smash into is not like glacial till, the unconsolidated, heterogeneous mixture of clay, sand, gravel, and boulders, dug-out, ground-out, and exposed by the force of a huge, slow, and inexorably moving glacier. Instead, this soil and rocky ground is exposed to the falling drops of snowmelt because of a combination of wind and the tiny, enduring force exerted by drops of water over season after season. The thunderous and earth-shaking meetings of the falling boulder and the surface below not only remove any lingering ground cover not previously removed by falling melt drops but also break and disintegrate the boulder into a couple of smaller 512–256-mm-sized boulders (this process will continue as headward erosion and bifurcation work in harmony to reduce a giant to sediment). The combination of falling melt drops and heavy boulder collisions below removes any possible semblance of soil or earthy flesh and exposes the underlying, intimate bones of the globe.

Gradually, the single drops increase to a small rush—they join to form a splashing, rebounding, helter-skelter cascade, many separate rivulets that trickle and sparkle in the bright sunlight and then run their way down the face of the granite mountain. A few of the smaller boulders continue their downward journey and also fall and run their way down the face of the mountain. At an indented ledge halfway down the mountain slope, a pool forms whose beauty, clarity, phantom blue eye, and sweet iciness provide the visitor with an incomprehensible, incomparable gift—a blessing from earth.

The mountain pool fills slowly, tranquil under the sunlit blue sky, reflecting the pines, snow, and sky around and above it, an open invitation to lie down and drink, and to peer into the glass-clear, deep waters, so clear that it seems possible to reach down over 50 feet and touch the very bowels of the mountain. The pool has no transition from shallow margin to depth; it is simply deep and pure. But the pool's depth is not permanent. Like all things in Nature, the pool changes—in the pool's case, it changes in size and depth but not because of the onslaught of melt water entry but because of the boulders that impact, now and then, with gigantic splashes, and then many of them sink to the bottom, raising the height of the floor level of the pool and decreasing the pool's depth, causing more snow melt water to pour over the edge.

As the pool fills with more melt water and the boulders settle in the depths and the pool's surface is glasslike again, we wish to freeze time, to hold this place and this pool in its perfect state forever; it is such a rarity to us in our modern world. But this cannot be—Mother Nature calls, prodding, urging, pushing—and for a brief instant, the water laps in the breeze against the outermost edge of the ridge, then a trickle flows over the rim. And then a smaller cobble (122 mm) teeter totters on the edge. Eventually, the giant hand of gravity reaches out and tips the overflowing melt and swaying cobble onward and they continue their downward journey, following the path of least resistance to their next destination, several thousand feet below.

When the overflow meets violently with the angled, broken rocks below, it bounces, bursts, and mists its way against steep, V-shaped walls that form a small, deep valley, carved out over time by water and the forces of earth, still high in altitude, but its rock-strewn bed bent downward, toward the sea. The smaller cobble, unlike the growing stream of melt water, is decreasing in size due to several violent collisions and is now coarse gravel (25 mm) as it accompanies the melt water downward.

Within the valley confines, the melt water has grown from drops to rivulets to a small mass of flowing water. Again, the former boulder is coarse gravel size. And they flow and tumble along, through what is at first a narrow opening, and then gain strength, speed, and power as the V-shaped valley widens to form a U-shape. But the journey still continues, as the water mass and cobble pick up speed and they spill and tumble over stream-bound massive boulders. And then the flow and transport slow again.

At a larger but shallower pool, waters from higher elevations have joined the main body, and our coarse gravel is now fine gravel (5 mm): from the hillsides, from crevices, from springs, from rills, from mountain creeks. At the influent pool sides, all appears peaceful, quiet, restful—but not far away, at the effluent end of the pool, gravity takes control again. The overflow of melt water and fine gravel are flung over the jagged lip and cascade downward several hundred feet, where the waterfall again brings its load to a violent, crushing, mist-filled meeting.

The water separates and joins again and again, forming a deep, furious, wild stream that calms gradually as it continues to flow over lands that are less steep. The gravel piggy-backs water flow and spins along in the violent ride. The waters widen into pools overhung by vegetation, surrounded by tall trees. The pure, crystalline waters have become progressively discolored on their downward journey, stained brown with humic acid and tannins and literally filled with suspended non-cohesive sediments; the once-pure stream is now muddy. The fine gravel is now very fine gravel (2 mm) but still non-cohesive, and it continues to roll, skid, and skip along the bed layer.

The mass divides and flows in different directions, over different landscapes. Small streams divert and flow off into open country. Different soils work to retain or speed up the waters, and in some places the waters spread out into shallow swamps, bogs, marshes, fens, or mires. Other streams pause long enough to fill deep depressions in the land and form lakes. While in these lakes, for a time the water remains and slows in its journey to the sea. But this is only a short-term pause because lakes are only a short-term resting place in the water cycle. The water will eventually move on by evaporation or seepage into groundwater, outlying to form another stream; move it will. The water flow's splintered, broken, crushed, and somewhat smoother and rounded sediments enter the lakes, and many of the larger pieces settle at the openings to lakes and ponds. Smaller pebbles spread out to deeper reaches, and sand and silt deposit in deeper areas closer to the stream outlet. A smaller amount of the rocky load continues to flow with the main flow as it reforms the river once again. These other portions of the water mass stay, and the remaining rocky load moves on with the main flow. The speed of the flow changes and increases to form another river, now a fourth-grade stream, which braids its way through the landscape, heading for the sea. As it continues its downward journey toward sea level, it changes speed again and slows. The river bottom changes from rock and stone, to silt and clay. Plants begin to grow, stems thicken, and leaves broaden. The river is now full of life and the nutrients needed to sustain life. But the river courses onward to its destination, where the flowing rich mass slows its last and finally spills into the sea.

What happened to the very fine gravel? Did it disappear? Was it spilled into the sea?

Well, yes and no. Some reached the sea as sand (1 mm).

The others?

Well, the others settled; they settled along, on, and in the riverbed.

Did they settle permanently . . . forever?

No; they did not.

The fine gravel and sand, remnants of the mighty boulder that fell from the mountain, are now settled, but not permanently. Remember, nothing in Nature is permanent. We are all guests. Short-term guests. And the residence of the settled sand (0.125 mm) is temporary. A combination of shear forces, a measurable amount of motion caused by critical shear stress, hyporheic exchange, bioturbation, groundwater upwelling, sediment resuspension, and benthic faunal activity all work alone or in harmony to lift and move the sediment to different locations. The inexorable fingering of erosion will convert sediment to a cohesive form, while other sediments will remain non-cohesive sediments; at least for the time being. In their new locations they will either rest, move on, and/or become contaminated sediments. With the passage of time and flow, these contaminated sediments will be bio-transformed. At least, that is our expectation . . . and our hope.

This is how it works. This is how mountains sometimes become sediments. This is how Nature sometimes works.

And here is an additional view of or look at the process of sediment formation. The end product produced in the following is much like the sediment produced by fluvial processes described previously, but water is not the main factor in this particular account.

EXAMPLE 2

If modern man were transported back in time, he would have instantly recognized the massive structure before him, even though he might have been taken aback at what he saw: a youthful mountain range with considerable mass, steep sides, and a height that certainly reached beyond any cloud. He would instantly relate to one particular peak—the tallest, most massive one. The polyhedron-shaped object, with its polygonal base and triangular faces culminating in a single sharp-tipped apex would have looked familiar—comparable in shape, though larger in size, to the largest of the Great Egyptian Pyramids, though the Pyramids were originally covered in a sheet of limestone, not the thick, perpetual sheet of solid ice and snow that covered the mountain peak.

But if man walked this same site in modern times, if he knew what had once stood upon this site, the changes would be obvious and startling—and entirely relative to time. What stood as an incomparable mountain peak eons ago, today man cannot see in its ancient majesty. In fact, he wouldn't give it a second thought as he walked across its remnants and through the vegetation that grows from its pulverized and amended remains.

Three hundred million years ago, the pyramid-shaped mountain peak stood in full, unchallenged splendor above the clouds, wrapped in a cloak of ice, a mighty fortress of stone, seemingly vulnerable to nothing, standing tallest of all—higher than any mountain ever stood—or will ever stand—on earth.

And so it stood, for millions upon millions of passings of the earth around the sun. Born when Mother Earth took a deep breath, the pyramid-shaped peak stood tall and undisturbed until millions of years later, when Mother Earth stretched. Today we would call this stretch a massive earthquake—humans have never witnessed one of such magnitude. Rather than registering on the Richter scale, it would have destroyed it.

But when this massive earthquake shattered the earth's surface, nothing we would call intelligent life lived on Earth—and it's a good thing.

During this massive upheaval, the peak shook to its very foundations, and after the initial shockwave and the hundred-plus aftershocks, the solid granite structure had fractured. This immense fracture was so massive that each aftershock widened it and loosened the base foundation of the pyramid-shaped peak itself. Only 10,000 years later (only a few seconds relative to geologic time), the fracture's effects totally altered the shape of the peak forever. During a horrendous windstorm, one of an intensity known only in earth's earliest days, a sharp tremor (emanating from deep within the earth and shooting up the spine of the mountain itself, up to the very peak) widened the gaping wound still more.

Decades of continued tremors and terrible windstorms passed (no present-day structure could withstand a blasting from such a wind), and finally, the highest peak of that time, of all time, fell. It broke off completely at its base and, following the laws of gravity (as effective and powerful a force then as today, of course), tumbled from its pinnacle position and fell more than 20,000 feet, straight down. It collided with the expanding base of the mountain range, the earth-shattering impact destroying several thousand acres. It finally came to rest (what remained intact) on a precipitous ledge at 15,000 feet in elevation. The pyramid-shaped peak, much smaller now, sat precariously perched on the precipitous ledge for about 5 million years.

Nothing, absolutely nothing, is safe from time. The most inexorable natural law is that of entropy. Time and entropy mean change and decay—harsh, sometimes brutal, but always inevitable. The bruised, scarred, truncated, but still massive rock form, once a majestic peak, was now a victim of Nature's way. Nature, with its chief ally, time, at its side, works to degrade anything and everything that has substance and form. For better or for worse, in doing so, Nature is ruthless, sometimes brutal, and always inevitable—but never without purpose.

While resting on the ledge, the giant rock, over the course of that 5 million years, was exposed to constantly changing conditions. For several thousand years, earth's climate was unusually warm—almost tropical—everywhere. Throughout this warm era, the rock was not covered with ice and snow but instead baked in intense heat; steamed in hot rain; and seared in the gritty, heavy windstorms that arose and released their abrasive fury, sculpting the rock's surface each day for more than 10,000 years.

Then came a pause in the endless windstorms and upheavals of the young planet, a span of time when the weather wasn't furnace-hot or arctic-cold but moderate. The rock was still exposed to sunlight, but at lower temperatures; to rainfall at increased levels; to fewer windstorms of increased fury. The climate remained so for some years—then the cycle repeated itself—arctic cold, moderately warm, furnace hot— and the cycle continued.

During the last of these cycles, the rock, considerably affected by physical and chemical exposure, was reduced in size and shape even more. Considerably smaller now than when it landed on the ledge (and a mere pebble compared to its former size), it fell again, 8,000 feet to the base of the mountain range, coming to rest on a bed of talus. Reduced in size still more, it remained on its sloping talus bed for many more thousand years.

Somewhere around 15,000 B.C., the rock form, continuously exposed to chemical and mechanical weathering, its physical structure weakened by its long-ago falls, fractured, split—broke into ever-decreasing-sized rocks, until the largest intact fragment left from the original rock was no bigger than a four-bedroom house. But change did not stop, and neither did time, rolling on until (about the time the Egyptians were building their pyramids) the rock was reduced, by this long, slow decaying process, to roughly 10 feet square.

Over the next thousand years, the rock continued to decrease in size, wearing, crumbling, flaking away, surrounded by fragments of its former self, until it was about the size of a beach ball. Covered with moss and lichen, a web of fissures, tiny crevices and fractures were now woven through the entire mass.

Over the next thousand or so years, via *bare rock succession*, what had once been the mother of all mountain peaks, the highest point on earth, had been reduced to nothing more than a handful of soil.

How did this happen? What is "bare rock succession?

If a layer of soil is completely stripped off land by natural means (water, wind, etc.), by anthropogenic means (tillage plus erosion), or by cataclysmic occurrence (a massive landslide or earthquake), only after many years can a soil-denuded area return to something approaching its original state or can a bare rock be converted to soil. But given enough time—perhaps a millennium—the scars heal over, and a new, virgin layer of soil forms where only bare rock once existed. We call the series of events that take place in this restoration process bare rock succession. It is indeed a true "succession"—with identifiable stages. Each stage in the pattern dooms the existent community as it succeeds the state that existed before.

Bare rock, however it is laid open to view, exposed to the atmosphere. The geologic processes that cause weathering begin, breaking down the surface into smaller and smaller fragments. Many forms of weathering exist, and all effectively reduce bare rock surface to smaller particles or chemicals in solution.

Lichens appear to cover the bare rock first. These hardy plants grow on the rock itself. They produce weak acids that assist in the slow weathering of the rock surface. The lichens also trap wind-carried soil particles, which eventually produce a very thin soil layer—a change in environmental conditions that gives rise to the next stage in bare rock succession.

Mosses replace lichens, growing in the meager soil the lichens and weathering provide. They produce a larger growing area and trap even more soil particles, providing a moister bare rock surface. The combination of more soil and moisture establishes abiotic conditions that favor the next succession stage.

Now the seeds of herbaceous plants invade what was once bare rock. Grasses and other flowering plants take hold. Organic matter provided by the dead plant tissue is added to the thin soil, while the rock still weathers from below. More and more organisms join the community as it becomes larger and more complex.

By this time, the plant and animal community is fairly complicated. The next major invasion is by weedy shrubs that can survive in the amount of soil and moisture present. As time passes, the process of building soil speeds up as more and more plants and animals invade the area. Soon trees take root and forest succession is evident. Many years are required, of course, before a climax forest will grow here, but the scene is set for that to occur.

Today, only the remnants of the former, incomparable pyramid-shaped peak are left. Sediment—sediment packed full of organic humus, sediment that looks like rocky road chocolate (pebbled mud) when wet and that, when dry, most people would think was just a handful of pebbles, sediment, or just plain old dirt. And that is what it is.

EXAMPLE 3

This chapter has detailed the May 18, 1980, eruption of Mt St Helens and several of the consequences of the cataclysmic event, including the loss of 57 known humans, more than 7,000 large mammals, millions of old-growth forest trees, and other trees, along with ground-hugging plants. In addition, the landscape was modified in dramatic fashion. Even now, 42 years later, the transformation not only of the mountain itself but of the northernmost landscape where the almost horizontal blast created a 14-mile kill zone are clearly evident (especially by helicopter flyover)—this is especially notable to those of us who hiked in the area before the eruption and several times after the eruption.

But there is more. There is always more whenever a cataclysmic event occurs. More but not necessarily noticeable—unless—unless you live downstream of streamflow, stream damage, riparian damage, aquatic habitat damage, and/or residential damage. Then the occurrence of an event like the eruption of Mt St Helens is evident—maybe to the nth degree. And this was and is certainly the case with the Mt St Helens eruption.

Downstream implications were made apparent due to the lahars (volcanic mudflows) that raced down river valleys, downstream, overtopping banks and flooding low-lying valleys. Sediment was the issue; it clogged channels in the Toutle River, the Cowlitz River, and eventually the Columbia River, 75 miles downstream from the volcano. In the Columbia River, the accumulation of sediment reduced the water depth from 40 feet to 14 feet, halting ship traffic, which caused severe economic distress to ports (USGS, 2020). In the Cowlitz and Toutle Rivers, accumulated channel sediment resulted in raised river levels and reduced the flood protection provided by local levees.

Years after the eruption, stream water flushed record-breaking amounts of sediments downstream. Long-term measurements by U.S. Geological Survey (USGS) show that the rate of sediment transport was among the highest in the world for several years.

What about today? Well, in 2020, when I sampled various locations along the Toutle and Cowlitz Rivers, I found, via filter sampling, still a large content of sediment, dissolved solids, and total dissolved solids (TDS) in the samples at rates of a few times greater than before the eruption (based on comparison with other USGS and water authority measurements pre-eruption).

The bottom line: these sediment loads continue to pose challenges to authorities responsible for flood control and fisheries. Truth be told, the streams have told us about the eruption of Mt St Helens in as glaring a fashion as the total devastation that occurred at the site.

NOTES

1. Source: USGS, Video Transcript Mount St. Helens: May 18, 1980). Accessed 02/15/2022 @ rmcclymont@usgs.gov (Public Domain).
2. Adaptation from F. Spellman's (2016) *Environmental Science*. Lanham, MD: Bernan Press.

REFERENCES

USGS (United States Geological Survey). (2020). *Mount St. Helens' 1980 Eruption.* Washington, DC: United States Geological Survey.

9 When the Land Subsides

INTRODUCTION

When the ground you stand on, build on, work on, and/or live on falls or sinks, and everything is destroyed or inundated with sea water, this is one of those events that drives humans and other lifeforms to migrate to greener pastures—solid ones. The sinking or falling of land refers to the downward shift of portions of Earth's surface—there are several causes of this phenomenon, and many are discussed in this and subsequent chapters.

To get even to the edge or the margin of understanding land subsidence, one must have some knowledge of the key substance involved: land (a.k.a. soil)—soil (a.k.a. land). So included in this chapter is a brief introduction to soil that is designed to provide a foundation for understanding land subsidence.

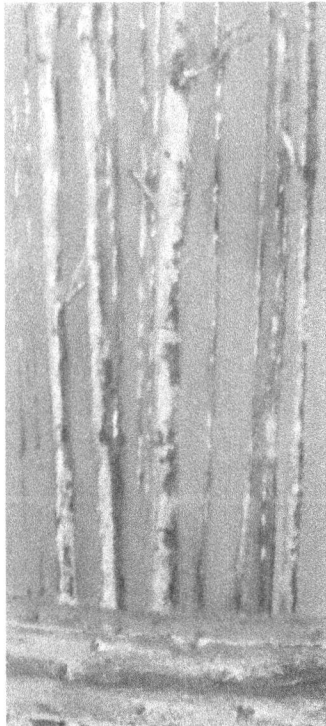

FIGURE 9.1 Stand of beech trees before subsidence.

Source: Artwork by F. Spellman (2020).

DOI: 10.1201/9781003295211-9

SOIL[1]

We take soil for granted. It's always been there, right?—with the implied corollary that it will always be there—right? But where does soil come from?

Of course, soil was formed, and in a never-ending process, it is still being formed. However, soil formation is a slow process—one at work over the course of millennia, as mountains are worn away to dust through bare rock succession.

Any activity, human or natural, that exposes rock to air begins the process. Through the agents of physical and chemical weathering, through extremes of heat and cold, through storms and earthquake and entropy, bare rock is gradually worn away. As its exterior structures are exposed and weakened, plant life appears to speed the process along.

Lichens cover the bare rock first, growing on the rock's surface, etching it with mild acids and collecting a thin film of soil that is trapped against the rock and clings to it. This changes the conditions of growth so much that the lichens can no longer survive and are replaced by mosses.

The mosses establish themselves in the soil that was trapped and enriched by the lichens and collect even more soil. They hold moisture to the surface of the rock, setting up another change in environmental conditions.

Well-established mosses hold enough soil to allow herbaceous plant seeds to invade the rock. Grasses and small flowering plants move in, sending out fine root systems that hold more soil and moisture, and work their way into minute fissures in the rock's surface. More organisms join the increasingly complex community.

Weedy shrubs are the next invaders, with heavier root systems that find their way into every crevice. Each stage of succession affects the decay of the rock's surface and adds its own organic material to the mix. Over the course of time, mountains are worn away, eaten away to soil, as time, plants, weather, and extremes of weather work on them.

The parent material, the rock, becomes smaller and weaker as the years, decades, centuries, and millennia go by, creating the rich, varied, and valuable mineral resource we call soil.

SOIL: WHAT IS IT?

Perhaps no term causes more confusion in communication between various groups of average persons, soil scientists, soil engineers, and earth scientists than the word "soil" itself. In simple terms, soil can be defined as the topmost layer of decomposed rock and organic matter, which usually contains air, moisture, and nutrients and can therefore support life. Most people would have little difficulty in understanding and accepting this simple definition. Then why are various groups confused on the exact meaning of the word soil? Quite simply, confusion reigns because soil is not simple—it is quite complex. In addition, the term soil has different meanings to different groups (like pollution, the exact definition of soil is a personal judgment call). Let's look at how some of these different groups view soil.

The average person seldom gives soil a first or second thought. Why should they?—soil isn't that big a deal—that important—it doesn't impact their lives, pay their bills, or feed the bulldog, right?

Not exactly. Not directly.

The average person seldom thinks about soil as soil. He or she may think of soil in terms of dirt, but hardly ever as soil. Why is this? Having said the obvious about the confusion between soil and dirt, let's clear up this confusion.

First, soil is not dirt. Dirt is misplaced soil—soil where we don't want it, contaminating our hands, clothes, automobiles, tracked in on the floor. Dirt we try to clean up and to keep out of our living environments.

Second, soil is too special to be called dirt. Why? Because soil is mysterious and, whether we realize it or not, essential to our existence. Because we think of it as common, we relegate soil to an ignoble position. As our usual course of action, we degrade it, abuse it, throw it away, contaminate it, ignore it—we treat it like dirt, and only feces hold a lowlier status. Soil deserves better.

Why?

Again, because soil is not dirt—how can it be? It is not filth, or grime, or squalor. Instead, soil is clay, air, water, sand, loam, organic detritus of former lifeforms, and most important, the amended fabric of Earth itself; if water is Earth's blood, and air is Earth's breath, then soil is its flesh, bone, and marrow—simply put, soil is the substance that most life depends on.

Soil scientists (or pedologists): a group interested in soils as a medium for plant growth. Their focus is on the upper meter or so beneath the land surface (this is known as the weathering zone, which contains the organic-rich material that supports plant growth) directly above the unconsolidated parent material. Soil scientists have developed a classification system for soils based on the physical, chemical, and biological properties that can be observed and measured in the soil.

Soils engineers: typically soils specialists who look at soil as a medium that can be excavated using tools. Soils engineers are not concerned with the plant-growing potential of a soil but rather are concerned with a particular soil's ability to support a load. They attempt to determine (through examination and testing) a soil's particle size, particle-size distribution, and plasticity.

Earth scientists (or geologists): have a view that typically falls between those of pedologists and soils engineers—they are interested in soils and the weathering processes as past indicators of climatic conditions and in relation to the geologic formation of useful materials ranging from clay deposits to metallic ores.

Would you like to gain new understanding of soil? Take yourself out to a plowed farm field somewhere. Reach down and pick up a handful of soil and look at it—really look at it closely. What are you holding in your hand? Look at the two descriptions of that handful of soil that follow, and you may gain a better understanding of what soil is and why it is critically important to us all.

1. A handful of soil is alive, a delicate living organism—as lively as an army of migrating caribou and as fascinating as a flock of egrets. Literally teeming with life of incomparable forms, soil deserves to be classified as an independent ecosystem or, more correctly stated, as many ecosystems as possible.
2. When we reach down and pick up a handful of soil, exposing Earth's stark bedrock surface, it should remind us (and maybe startle some of us) that without its thin living soil layer, Earth is a planet as lifeless as our own moon (Spellman and Whiting, 2006).

   The purpose of this chapter is to provide information: information on soil—on just about every aspect of soil. And in so doing, we do not lose sight of the fact that we currently have a problem with soil. The problem? Consider, for example, that in the United States, along with almost every other country, we are experiencing a continuing increase in the amount and type of materials we discard—as we stated earlier, we have become a throwaway society. This places a significant burden on disposal sites—landfills. It may surprise you to know that in the past three decades, a significant amount of the material being landfilled is contaminated soil. This contaminated soil is only part of the problem, however. Remember, soil is in direct contact with water and air (the atmosphere). Since water, air, and soil exist in a complex interrelationship, this contact causes problems with water and soil as well.

   In the past, soil pollution problems have been exacerbated by the physical, economic, and technical limitations associated with the technologies currently available for the remediation of contaminated soil. Progress has been made in this vital area, however. While problems are still to be solved in the areas of soil remediation, technology is not necessarily one of them.

DEFINITIONS OF KEY TERMS

Every branch of science, including soil science, has its own language. The terminology used herein is as different from that of astrophysics as astronomy is from botany. To work even at the edge of soil science, soil pollution, and soil pollution remediation, you must acquire a familiarity with the vocabulary used in this text.

Definitions

ablation till: a super-glacial coarse-grained sediment or till, accumulating as the sub-adjacent ice melts and drains away, finally deposited on the exhumed sub-glacial surface.

absorption: movement of ions and water into plant roots because of either metabolic processes by the root (active absorption) or as a result of diffusion along a gradient (passive absorption).

acid rain: atmospheric precipitation with pH values less than about 5.6, the acidity being due to inorganic acids such as nitric and sulfuric that are formed when oxides of nitrogen and sulfur are emitted into the atmosphere.

acid soil: a soil with a pH value of <7.0 or neutral. Soils may be naturally acid from their rocky origin, by leaching, or may become acid from decaying leaves or from soil additives such as aluminum sulfate (alum). Acid soils can be neutralized by the addition of lime products.

actinomycetes: a group of organisms intermediate between the bacteria and the true fungi that usually produce a characteristic branched mycelium. Includes many (but not all) organisms belonging to the order of Actinomycetales.

adhesion: molecular attraction that holds the surfaces of two substances (e.g., water and sand particles) in contact.

adsorption: the attraction of ions or compounds to the surface of a solid.

aeration, soil: the process by which air in the soil is replaced by air from the atmosphere. In a well-aerated soil, the soil air is similar in composition to the atmosphere above the soil. Poorly aerated soils usually contain more carbon dioxide and correspondingly less oxygen than the atmosphere above the soil.

aerobic: growing only in the presence of molecular oxygen, as aerobic organisms.

aggregate, soil: soil structural units of various shapes, composed of mineral and organic material, formed by natural processes and having a range of stabilities.

agronomy: a specialization of agriculture concerned with the theory and practice of field crop production and soil management. The scientific management of land.

air capacity: percentage of soil volume occupied by air spaces or pores.

air porosity: the proportion of the bulk volume of soil that is filled with air at any given time or under a given condition, such as a specified moisture potential, usually the large pores.

alkali: a substance capable of liberating hydroxide ions in water, measured by a pH of more than 7.0 and possessing caustic properties; it can neutralize hydrogen ions, with which it reacts to form salt and water, and is an important agent in rock weathering.

alluvium: a general term for unconsolidated, granular sediments deposited by rivers.

amendment, soil: any substance other than fertilizers (such as compost, sulfur, gypsum, lime, and sawdust) used to alter the chemical or physical properties of a soil, generally to make it more productive.

ammonification: the production of ammonia and ammonium-nitrogen through the decomposition of organic nitrogen compounds in soil organic matter.

anaerobic: without molecular oxygen.

anion: an atom which has gained one or more negatively charged electrons and is thus itself negatively charged.

aspect (of slopes): the direction that a slope faces with respect to the sun.

assimilation: the taking up of plant nutrients and their transformation into actual plant tissues.

Atterburg limits: water contents of fine-grained soils at different states of consistency.

autotrophs: plants and microorganisms capable of synthesizing organic compounds from inorganic materials by either photosynthesis or oxidation reactions.

available water: the portion of water in a soil that can be readily absorbed by plant roots. The amount of water released between the field capacity and the permanent wilting point.

bedrock: the solid rock underlying soils and the regolith in depths ranging from zero (where exposed by erosion) to several hundred feet.

biological function: the role played by a chemical compound or a system of chemical compounds in living organisms.

biomass: the total weight of living biological organisms within a specified unit (area, community, population).

biome: a major ecological community extending over large areas.

blow-out: a deflation depression, eroded by wind from the face of a vegetated dune.

breccia: a rock composed of coarse angular fragments cemented together.

calcareous soil: containing sufficient calcium carbonate (often with magnesium carbonate) to effervesce visibly when treated with hydrochloric acid.

caliche: a layer near the surface, cemented by secondary carbonates of calcium or magnesium precipitated from the soil solution. It may occur as a soft, thin soil horizon, as a hard, thick bed just beneath the solum, or as a surface layer exposed by erosion.

capillary water: held within the capillary pores of soils, mostly available to plants.

catena: the sequences of soils which occupy a slope transect, from the topographic divide to the bottom of the adjacent valley.

cation: an atom which has lost one or more negatively charged electrons and is thus itself positively charged.

Chelate: (Greek, claw) a complex organic compound containing a central metallic ion surrounded by organic chemical groups.

class, soil: a group of soils having a definite range in a particular property such as acidity, degree of slope, texture, structure, land-use capability, degree of erosion, or drainage.

clay: a soil separate consisting of particles <0.0002 mm in equivalent diameter.

cohesion: holding together: force holding a solid or liquid together, owing to attraction between like molecules. Decreases with rise in temperature.

colloidal: matter of very fine particle size.

convection: a process of heat transfer in a fluid involving the movement of substantial volumes of the fluid concerned. Convection is very important in the atmosphere and, to a lesser extent, in the oceans.

denitrification: the biochemical reduction of nitrate or nitrite to gaseous nitrogen, either as molecular nitrogen or as an oxide of nitrogen.

detritus: debris from dead plants and animals.

diffusion: the movement of atoms in a gaseous mixture, or ions in a solution, primarily because of their own random motion.

drainage: the removal of excess water, both surface and subsurface, from plants. All plants (except aquatics) will die if exposed to an excess of water.

duff: the matted, partly decomposed organic surface layer of forest soils.

erosion: the wearing away of the land surface by running water, wind, ice, or other geological agents, including such processes as gravitational creep.

eutrophication: a process of lake aging whereby aquatic plants are abundant and waters are deficient in oxygen. The process is usually accelerated by enrichment of waters with surface runoff containing nitrogen and phosphorus.

evapotranspiration: the combined loss of water from a given area, during a specified period, by evaporation from the soil surface and by transpiration from plants.

exfoliation: mechanical or physical weathering that involves the disintegration and removal of successive layers of rock mass.

fertility, soil: the quality of a soil that enables it to provide essential chemical elements in quantities and proportions for the growth of specified plants.

fixation: the transformation in soil of a plant nutrient from an available to an unavailable state.

fluvial: deposits of parent materials laid down by rivers or streams.

friable: a soil consistency term pertaining to the ease of crumbling of soils.

heaving: the partial lifting of plants, buildings, roadways, and fence posts, and so on out of the ground, because of freezing and thawing of the surface soil during the winter.

heterotroph: an organism capable of deriving energy for life processes only from the decomposition of organic compounds and incapable of using inorganic compounds as sole sources of energy or for organic synthesis.

horizon, soil: a layer of soil, approximately parallel to the soil surface, differing in properties and characteristics from adjacent layers below or above it.

humus: stable fraction of the soil organic matter (usually dark in color) remaining after the major portions of added plant and animal residues have decomposed.

hydration: the incorporation of water into the chemical composition of a mineral, converting it from an anhydrous to a hydrous form; the term is also applied to a form of weathering in which hydration swelling creates tensile stress within a rock mass.

hydraulic conductivity: the rate at which water can move through a soil.

hydrolysis: the reaction between water and a compound (commonly a salt). The hydroxyl from the water combines with the anion from the compound undergoing hydrolysis to form a base; the hydrogen ion from the water combines with the cation from the compound to form an acid.

hygroscopic coefficient: the amount of moisture in a dry soil when it is in equilibrium with some standard relative humidity nears a saturated atmosphere (about 98 percent), expressed in terms of percentage based on oven-dry soil.

infiltration: the downward entry of water into the soil.

ions: atoms which have lost or gained one or more negatively charged electrons.

land classification: the arrangement of land units into various categories based upon the properties of the land and its suitability for some particular purpose.

leaching: the removal of materials in solution from the soil by percolating waters.

Liebig's law: the growth and reproduction of an organism are determined by the nutrient substance (oxygen, carbon dioxide, calcium, etc.) that is available in minimum quantity with respect to organic needs, the limiting factor.

loam: the textural-class name for soil having moderate amounts of sand, silt, and clay.

loess: an accumulation of wind-blown dust (silt) which may have undergone mild digenesis.

marl: an earthy deposit consisting mainly of calcium carbonate, usually mixed with clay. Marl is used for liming acid soils. It is slower acting than most lime products used for this purpose.

mineralization: the conversion of an element from an organic form to an inorganic state because of microbial decomposition.

nitrogen fixation: the biological conversion of elemental nitrogen (N_2) to organic combinations, or to forms readily utilized in biological processes.

osmosis: the movement of a liquid across a membrane from a region of high concentration to a region of low concentration. Water and nutrients move into roots independently.

oxidation: the loss of electrons by a substance.

parent material: the unconsolidated and chemically weathered mineral or organic matter from which the solum of soils is developed by pedogenic processes.

ped: a unit of soil structure such as an aggregate, crumb, prism, block, or granule, formed by natural processes.

pedogenic/pedagogical process: any process associated with the formation and development of soil.

pH: the degree of acidity or alkalinity of the soil. Also referred to as soil reaction, this measurement is based on the pH scale where 7.0 is neutral (values from 0.0 to 7.0 are acid and values from 7.0 to 14.0 are alkaline). The pH of soil is determined by a simple chemical test where a sensitive indicator solution is added directly to a soil sample in a test tube.

photosynthesis: the process by which green leaves of plants, in the presence of sunlight, manufacture their own needed materials from carbon dioxide in the air and water and minerals taken from the soil.

porosity, soil: the volume percentage of the total bulk not occupied by solid particles.

profile, soil: a vertical section of the soil through all its horizons and extending into the parent material.

reduction: the gain of electrons, and therefore the loss of positive valence charge by a substance.

regolith: the unconsolidated mantle of weathered rock and soil material on the earth's surface; loose earth materials above solid rock.

rock: the material that forms the essential part of the earth's solid crust, including loose incoherent masses such as sand and gravel, as well as solid masses of granite and limestone.

rock cycle: the global geological cycling of lithospheric and crustal rocks from their igneous origins through all any stages of alteration, deformation, resorption, and reformation.

runoff: the portion of the precipitation on an area that is discharged from the area through stream channels.

salinization: the process of accumulation of salts in soil.

sand: a soil particle between 0.05 and 2.0 mm in diameter; a soil textural class.

silt: a soil separate consisting of particles between 0.05 and 0.002 mm in equivalent diameter. A soil textural class.

slope: the degree of deviation of a surface from horizontal, measured in a numerical ratio, percent, or degrees.

soil: an assemblage of loose and normally stratified granular minerogenic and biogenic debris at the land surface; it is the supporting medium for the growth of plants.

soil air: the soil atmosphere; the gaseous phase of the soil, being that volume not occupied by soil or liquid.

soil horizon: a layer of soil, approximately parallel to the soil surface, with distinct characteristics produced by soil-forming processes. These characteristics form the basis for systematic classification of soils.

soil profile: a vertical section of the soil from the surface through all its horizons, including C horizons.

soil structure: the combination or arrangement of primary soil particles into secondary particles, units, or peds. These secondary units may be, but usually are not, arranged in the profile in such a manner as to give a distinctive characteristic pattern. The secondary units are characterized and classified based on size, shape, and degree of distinctness into classes, types, and grades, respectively.

soil texture: the relative proportion of the various soil separates in a soil.

soluble: will dissolve easily in water.

solum: (plural sola) the upper and most weathered part of the soil profile; the A, E, and B horizons.

subsoil: that part of the soil below the plow layer.

till: unstratified glacial drift deposited directly by the ice and consisting of clay, sand, gravel, and boulders intermingled in any proportion.

tilth: the physical condition of soil as related to its ease of tillage, fitness as a seedbed, and its impedance to seedling emergence and root penetration.

topsoil: the layer of soil moved in cultivation.

weathering: all physical and chemical changes produced in rocks, at or near the earth's surface, by atmospheric agents.

ALL ABOUT SOIL

Before we begin a journey that takes us through the territory that is soil and examine soil from micro to macro levels, we need to stop for a moment and discuss why, beyond the obvious reason, soil is so important to us—to our environment—to our very survival. Is soil really that big a deal? Is it that important? Do we need to even think about soil?

Yes, yes, and yes. Soil is all these things and more.

Functions of Soil

We normally think of or relate soil to our backyards, to farms, to forests, or to a regional watershed. We think of soil as the substance upon which plants grow. This is generally about as far as the average person's thoughts extend concerning soil and its usefulness to us. Soils play other roles, though. They have five main functions important to us: (1) soil is a medium for plant growth; (2) soils regulate our water supplies;

(3) soils are recyclers of raw materials; (4) soils provide a habitat for organisms; and (5) soils are used as an engineering medium.

Let's take a closer look at each of the functions of soils.

Soil: A Plant Growth Medium

We are all aware of the primary function of soil: soil serves as a plant growth medium; it is responsible for plant growth (a function that becomes more important with each passing day as Earth's population continues to grow). However, while it is true that soil is a medium for plant growth (thus critical to maintaining life), let us also point out that soil is alive as well. Soil exists in paradox: we depend on soil for life, and at the same time, soil depends on life; its very origin, its maintenance, and its true nature are intimately tied to living plants and animals. What does this mean?

As a plant growth medium, soil provides vital resources and performs important functions for the plant. To grow in soil, plants must have water and nutrients—soil provides these. To grow and to sustain growth, a plant must have a root system—soil provides pore spaces for roots. To grow and maintain growth, a plant's roots must have oxygen for respiration and carbon dioxide exchange and ultimate diffusion out of the soil—soil provides the air and pore spaces (the soil's ventilation system) for this. To continue to grow, a plant must have support—soil provides this support.

In order for a seed to grow, it must be planted in soil and exposed to the proper amount of sunlight for growth to occur. The soil must provide nutrients through a root system that has space to grow, as well as a continuous stream of water (it requires about 500 grams of water to produce 1 g of dry plant material) for root nutrient transport and plant cooling and a pathway for both oxygen and carbon dioxide transfer. Just as importantly, soil water provides the plant with its normal fullness or tension (turgor) it needs to stand—the structural support it needs to face the sun for photosynthesis to occur.

As well as the functions stated previously, soil is also an important moderator of temperature fluctuations. If you have ever dug in a garden on a hot summer day, you probably noticed that the soil was warmer (even hot) on the surface, but much cooler just a few inches below the surface.

Soil: Regulator of Water Supplies

When we walk on land, few of us probably realize that we are walking across a bridge. This bridge (in many areas) transports us across a veritable ocean of water below us, deep—or not so deep—under the surface of the earth.

Consider what happens to rain. Where does the rainwater go? Some, falling directly over water bodies, become part of the water body again. But an enormous amount falls on land. Where does it go? Some of the water, obviously, runs off—always following the path of least resistance. In modern communities, stormwater runoff is a hot topic. Cities have taken giant steps to try to control runoff—to send it where it can be properly handled, to prevent flooding.

Let's take a closer look at precipitation and the "sinks" it "pours" into, and then relate this usually natural operation to soil water. We begin with surface water, and then move on to that ocean of water below the soil's surface: groundwater.

Surface water (water on the earth's surface, as opposed to subsurface water—groundwater) is mostly a product of precipitation—rain, snow, sleet, or hail. Surface water is exposed or open to the atmosphere and results from the movement of water on and just under the earth's surface (overland flow). This overland flow is the same thing as surface runoff, which is the amount of rainfall that passes over the earth's surface. Specific sources of surface water include rivers, streams, lakes, impoundments, shallow wells, rain catchments, and tundra ponds or meskegs (peat bogs).

Most surface water is the result of surface runoff. The amount and flow rate of surface runoff is highly variable. This variability stems from two main factors: (1) human interference and (2) natural conditions. In some cases, surface water runs quickly off land. Generally, this is undesirable (from a water resources standpoint) because it does not provide enough time for water to infiltrate into the ground and recharge groundwater aquifers. Other problems associated with quick surface water runoff are erosion and flooding. Probably the only good thing that can be said about surface water that quickly runs off land is that it does not have enough time (normally) to become contaminated with high mineral content. Surface water running slowly off land may be expected to have all the opposite effects.

Surface water travels over the land to what amounts to a predetermined destination. What factors influence how surface water moves? Surface water's journey over the face of the earth typically begins at its drainage basin, sometimes referred to as its drainage area, catchment, and/or watershed. For a groundwater source, this is known as the recharge area—the area from which precipitation flows into an underground water source.

A surface water drainage basin is usually an area measured in square miles, acres, or sections. If a city takes water from a surface water source, the size of (and what lies within) the drainage basin is essential information for the assessment of water quality.

We all know that water doesn't run uphill. Instead, surface water runoff (like the flow of electricity) follows the path of least resistance. Water within a drainage basin will naturally (by the geological formation of the area) be shunted toward one primary watercourse (a river, stream, creek, or brook) unless some manmade distribution system diverts the flow.

Various factors directly influence the surface water's flow over land. The principal factors are:

Rainfall Duration: The length of the rainstorm affects the amount of runoff. Even a light, gentle rain will eventually saturate the soil if it lasts long enough. Once the saturated soil can absorb no more water, rainfall builds up on the surface and begins to flow as runoff.

Rainfall Intensity: The harder and faster it rains, the more quickly soil becomes saturated. With hard rains, the surface inches of soil quickly become inundated. With short, hard storms, most of the rainfall may end up as surface runoff, because the moisture is carried away before significant amounts of water are absorbed into the earth.

Soil Moisture: Obviously, if the soil is already laden with water from previous rains, the saturation point will be reached sooner than if the soil was dry. Frozen soil also inhibits water absorption: up to 100 percent of snow melt or rainfall on frozen soil will end up as runoff because frozen ground is impervious.

Soil Composition: Runoff amount is directly affected by soil composition. Hard rock surfaces will shed all rainfall, obviously, but so will soils with heavy clay composition. Clay soils possess small void spaces that swell when wet. When the void spaces close, they form a barrier that does not allow additional absorption or infiltration. On the opposite end of the spectrum, coarse sand allows easy water flow-through, even in a torrential downpour.

Vegetation Cover: Runoff is limited by ground cover. Roots of vegetation and pine needles, pinecones, leaves, and branches create a porous layer (sheet of decaying natural organic substances) above the soil. This porous "organic" sheet (ground cover) readily allows water into the soil. Vegetation and organic waste also act as a cover to protect the soil from hard, driving rains. Hard rains can compact bare soils, close off void spaces, and increase runoff. Vegetation and ground cover work to maintain the soil's infiltration and water-holding capacity. Note that vegetation and groundcover also reduce evaporation of soil moisture.

Ground Slope: Flat land water flow is usually so slow that large amounts of rainfall can infiltrate the ground. Gravity works against infiltration on steeply sloping ground where up to 80 percent of rainfall may become surface runoff.

Human Influences: Various human activities have a definite impact on surface water runoff. Most human activities tend to increase the rate of water flow. For example, canals and ditches are usually constructed to provide steady flow, and agricultural activities generally remove ground cover that would work to retard the runoff rate. On the opposite extreme, human-made dams are generally built to retard the flow of runoff.

Human habitations, with their paved streets, tarmac, paved parking lots, and buildings, create surface runoff potential, since so many surfaces are impervious to infiltration. All these surfaces hasten the flow of water, and they also increase the possibility of flooding, often with devastating results. Because of urban increases in runoff, a whole new field (industry) has developed: stormwater management.

Paving over natural surface acreage has another serious side effect. Without enough area available for water to infiltrate the ground and percolate through the soil to eventually reach and replenish—recharge—groundwater sources, those sources may eventually fail, with devastating impact on local water supply.

Now let's shift gears and look at groundwater.

Water falling to the ground as precipitation normally follows three courses. Some runs off directly to rivers and streams, some infiltrates to ground reservoirs, and the rest evaporates or transpires through vegetation. The water in the ground (groundwater) is "invisible" and may be thought of as a temporary natural reservoir. Almost all groundwater is in constant motion toward rivers or other surface water bodies.

Groundwater is defined as water below the earth's crust but above a depth of 2,500 feet. Thus, if water is located between the earth's crust and the 2,500-foot level, it is considered usable (potable) fresh water. In the United States, it is estimated "that at least 50% of total available freshwater storage is in underground aquifers" (Kemmer, 1979, p. 17).

In this text, we are concerned with that amount of water retained in the soil to ensure plant life and growth. Having said this, recall that earlier we stated that producing 1 g of dry plant material requires about 500 grams of water. Note that about 5 grams of this water becomes an integral part of the plant. Unless rainfall is frequent, you don't have to be a rocket scientist to figure out that the ability of soil to hold water against the force of gravity is very important—thus, one of the vital functions of soil is to regulate the water supply to plants.

SOILS: RECYCLERS OF RAW MATERIALS

Can you imagine what it would be like to step out into the open air and be hit by a stench that you could not only smell, but could almost reach out and grab you (like the situation we had in the cave earlier—but worse)? You look out upon the cluttered fields in front of your domicile and see nothing but stack upon stack upon stack of the sources of this horrible, putrefied, gagging stench. We are talking about plant and animal remains and waste (mountains of it), reaching toward the sky. "Impossible," you say. Well, thankfully (in most cases), you are right. However, if it were not for the power of the soil to recycle waste products, then this scene or something like it is imaginable and even possible—but of course it would be impossible, because there would be no life to die and to stack up anywhere, for that matter.

Soil is a recycler—probably the premier recycler on Earth. The simple fact is that if it were not for soil's incredible recycling ability, plants and animals would have run out of nourishment long ago. Soil recycles in other ways. For example, consider the geochemical cycles (i.e., the chemical interactions between soil, water, air, and life on earth) in which soil plays a significant role.

Soil possesses the incomparable ability and capacity to assimilate great quantities of organic wastes and turn them into beneficial organic matter (humus), then to convert the nutrients in the wastes to forms that can be utilized by plants and animals. In turn, the soil returns carbon to the atmosphere as carbon dioxide, where it again will eventually become part of living organisms through photosynthesis. Soil performs several different recycling functions—most of them good, some of them not so good.

Not so good? Yes. Consider one recycling function of soil that may not be so good. Soils have the capacity to accumulate substantial amounts of carbon as soil organic matter, which can have a major impact on global change such as greenhouse effect.

SOIL: HABITAT FOR SOIL ORGANISMS

One thing is certain: most soils are not dead and sterile things. The fact is a handful of soil is an ecosystem. It may contain up to billions of organisms, belonging to thousands of species. Let's take a look at Table 9.1, which lists a few (very few) of these organisms.

TABLE 9.1
Soil Organisms

Soil Organisms (A Representative Sample)

Microorganisms (protists)
    Bacteria
    Fungi
    Actinomycetes
    Algae
    Protozoa
Nonarthropod animals
    Nematodes
    Earthworms and potworms
Arthropod animals
    Springtails
    Mites
    Millipedes and centipedes
    Harvestman
    Ants
    Diplopoda
    Diptera
    Crustacea
Vertebrates
    Mice, moles, voles
    Rabbits, gophers, squirrels

Obviously, communities of living organisms inhabit the soil. What is not so obvious is that they are as complex and intrinsically valuable as are those organisms that roam the land surface and waters of Earth.

SOIL: AN ENGINEERING MEDIUM

We usually think of soil as being firm and solid—"solid ground" . . . terra firma. As solid ground (when it is), soil is usually a good substrate upon which to build highways and structures. We say "usually" because not all soils are firm and solid—some are not as stable as others. While construction of buildings and highways may be suitable in one location on one type of soil, it may be unsuitable in another location with different soil. To construct structurally sound, stable (and therefore reliable) highways and buildings, construction on soils and with soil materials requires knowledge of the diversity of soil properties—which really means that a knowledge of the engineering properties of soils is required.

Note that working with manufactured building materials that have been "engineered" to withstand certain stresses and forces is much different than working with natural soil materials, even though engineers have the same concerns about soils as they do with human-made building materials (concrete and steel). It is much more difficult to make these predictions or determinations for soil's ability to resist compression, to remain in place, its bearing strength, shear strength, and stability than it is to make the same determinations for manufactured building materials.

SOIL BASICS

Any fundamental discussion about soil should begin with a definition of what soil is. The word soil is derived through Old French from the Latin *solum*, which means floor or ground. A more concise definition is made difficult by the great diversity of soils throughout the globe. However, here is a generalized definition from the Soil Science Society of America:

Soil is unconsolidated mineral matter on the surface of the earth that has been subjected to and influenced by genetic and environmental factors of parent material, climate, macro- and microorganisms, and topography, all acting over a period of time and producing a product—soil—that differs from the material from which it is derived in many physical, chemical, and biological properties, and characteristics.

Engineers might define soil by saying that soil occupies the unconsolidated mantle of weathered rock making up the loose materials on the Earth's surface, commonly known as the regolith.

Soil can be described as a three-phase system, composed of a solid, liquid, and gaseous phase.

NOTE TO READERS

This phase relationship is important in dealing with soil pollution, because each of the three phases of soil is in equilibrium with the atmosphere, and with rivers, lakes, and the oceans. Thus, the fate and transport of pollutants are influenced by each of these components.

Soil is also commonly described as a mixture of air, water, mineral matter, and organic matter; the relative proportions of these four components greatly influence the productivity of soils. The interface (where the regolith meets the atmosphere) of these materials that make up soil is what concerns us here.

Keep in mind that the four major ingredients that make up soil are not mixed or blended like cake batter. Instead, pore spaces (vital to air and water circulation, providing space for roots to grow and microscopic organisms to live) are a major (and critically important) constituent of soil. Without sufficient pore space, soil would be too compacted to be productive. Ideally, the pore space will be divided roughly equally between water and air, with about one-quarter of the soil volume consisting of air and one-quarter consisting of water. The relative proportions of air and water in a soil typically fluctuate significantly as water is added and lost. Compared to surface soils, subsoils tend to contain less total pore space, less organic matter, and a larger proportion of micropores, which tend to be filled with water.

Let's take a closer look at the four major components (air, water, mineral matter, and organic matter) that make up soil.

Soil air circulates through soil pores in the same way air circulates through a ventilation system. Only when the pores (the ventilation ducts) become blocked by water or other substances does the air fail to circulate. Though soil pores normally interface with the atmosphere, soil air is not the same as atmospheric air. It differs in composition from place to place. Soil air also normally has a higher moisture content than the atmosphere. The content of carbon dioxide (CO_2) is usually higher as well and that of oxygen (O_2) lower than accumulations of these gases found in the atmosphere.

Earlier we stated that only when soil pores are occupied by water or other substances does air fail to circulate in the soil. For proper plant growth, this is of importance, because in soil pore spaces that are water dominated, air oxygen content is low and carbon dioxide levels are high, which restricts plant growth.

The presence of water in soil (often reflective of climatic factors) is essential for the survival and growth of plants and other soil organisms. Soil moisture is a major determinant of the productivity of terrestrial ecosystems and agricultural systems. Water moving through soil materials is a major force behind soil formation. Along with air, water, and dissolved nutrients, soil moisture is critical to the quality and quantity of local and regional water resources.

Mineral matter varies in size and is a major constituent of non-organic soils. Mineral matter consists of large particles (rock fragments) including stones, gravel, and coarse sand. Many of the smaller mineral matter components are made of a single mineral. Minerals in the soil (for plant life) are the primary source of most of the chemical elements essential for plant growth.

Soil organic matter consists primarily of living organisms and the remains of plants, animals, and microorganisms that are continuously broken down (biodegraded) in the soil into new substances that are synthesized by other microorganisms. These other microorganisms continually use this organic matter and reduce it to carbon dioxide (via respiration) until it is depleted, making repeated additions of new plant and animal residues necessary to maintain soil organic matter).

Now that we have defined soil, let's take a closer look at a few of the basics pertaining to soil and some of the common terms used in any discussion related to soil basics.

Soil is the layer of bonded particles of sand, silt, and clay that covers the land surface of the earth. Most soils develop in multiple layers. The topmost layer (topsoil) is the layer of soil moved in cultivation and in which plants grow. This topmost layer is an ecosystem composed of both biotic and abiotic components—inorganic chemicals, air, water, decaying organic material that provides vital nutrients for plant photosynthesis, and living organisms. Below the topmost layer is the subsoil (the part of the soil below the plow level, usually no more than a meter in thickness). Subsoil is much less productive, partly because it contains much less organic matter. Below that is the parent material, the unconsolidated (and chemically weathered) bedrock or other geologic material from which the soil is ultimately formed. The general rule of thumb is that it takes about 30 years to form 1 inch of topsoil from subsoil; it takes much longer than that for subsoil to be formed from parent material—the length of time depending on the nature of the underlying matter (Franck and Brownstone, 1992).

THE WATER CYCLE (HYDROLOGIC CYCLE)

Water is never stationary; it is constantly in motion. This phenomenon occurs because of the water or hydrologic cycle. In simple terms, the water cycle can be explained as follows: The sun helps transfer water from lakes and oceans to the land. As the sun shines on the earth, the surface water is heated and evaporates, forming an invisible gas that mixes with the air. This gas is water vapor; it is pure water without any minerals or bacteria in it. Water vapor rises in the air, then cools, and condenses into tiny drops of water that form clouds. Further cooling may form drops large enough to fall as rain. In this way, water is brought from the oceans to the land, where it reappears in springs and wells, soaks into the ground, or runs off again through streams and rivers back to the ocean. Of course, the actual movement of water on earth is much more complex. Three different methods of transport are involved in this water movement: evaporation, precipitation, and run-off.

Evaporation of water is a major factor in hydrologic systems. Evaporation is a function of temperature, wind velocity, and relative humidity. Evaporation (or vaporization) is, as the name implies, the formation of vapor. Dissolved constituents (such as salts) remain behind when water evaporates. Evaporation of the surface water of oceans provides most water vapor, though water can also vaporize through plants, especially from leaf surfaces. This process is evapotranspiration. Ice can also vaporize without melting first. However, this sublimation process is slower than vaporization of liquid water.

Precipitation includes all forms in which atmospheric moisture descends to earth—rain, snow, sleet, and hail. Before precipitation can occur, the water that enters the atmosphere by vaporization must first condense into liquid (clouds and rain) or solid (snow, sleet, and hail) before it can fall. This vaporization process absorbs energy, which is released in the form of heat when the water vapor condenses. You can best understand this phenomenon when you compare it to what occurs when water evaporates from your skin; this absorbs heat, making you feel cold. Note: The annual evaporation from ocean and land areas is the same as the annual precipitation.

Runoff is the flow back to the oceans of the precipitation that falls on land. This journey to the oceans is not always unobstructed. The flow back may be intercepted by vegetation (from which it later evaporates), a portion may be held in depressions, and some may infiltrate into the ground. A part of the infiltrated water is taken up by plant life and returned to the atmosphere through evapotranspiration, while the remainder either moves through the ground or is held by capillary action. Eventually, water drips, seeps, and flows its way back into lakes, ponds, rivers, streams, and the oceans.

SOIL WATER

Have you ever wondered what happens to water after it enters the soil? For the average person, probably not, but if you are to work in the soil science field, the answer to this question is one that you definitely need not only to know but must also have a full and complete understanding of. Water that enters the soil has (in simple terms) four ways it may go:

1. It may move on through the soil and percolate out of the root zone, where it may eventually reach the water table.
2. It may be drawn back to the surface and evaporate.

3. It may be taken up (transpired—used) by plants.
4. Finally, it may be "saved" in storage in the water profile.

What determines how much water ends up in each of these categories? It depends. Climate and the properties of the soil and the requirements of the plants growing in that soil all have an impact on how much water ends up in each of the categories. But don't forget the influence of anthropogenic actions (what we like to call the heavy hand of man)—people alter the movement of water not only by irrigation and stream diversion practices and by building but also by choosing which crops to plant and the types of tillage practices employed.

Land falls when land subsidence occurs. Land subsidence is the loss of surface elevation due to removal of subsurface support—this occurs in nearly every region of the United States. Subsidence occurs in many forms; it is quite diverse. The ground failures range from small local collapses to broad regional lowering of the earth's surface. According to Alice S. Allen of the U.S. Bureau of Mines, the Department of Interior describes land subsidence as being "merely the surface symptom, and the last step, of a variety of subsurface variety of subsurface mechanisms." The mechanisms (mostly due to human activities) of subsidence are diverse, and not all of them are fully understood. The mechanisms that are recognized or apparent to us include oxidation (dewatering) of organic peat soils, dissolution in limestone aquifers, first-time wetting of moisture-deficient low-density soils (hydrocompaction—the process of volume decrease and density increase that occurs when certain moisture-deficient deposits compact as they are wetted for the first time since burial). The vertical downward movement of the land surface that results from this process has also been termed "shallow subsidence" and "near-surface subsidence," natural compaction, liquefaction, crustal deformation, subterranean mining, and withdrawal of fluids (groundwater, petroleum, geothermal). Again, although we can list these mechanisms of subsidence, not all of them are fully understood. This is the case, of course, because land subsidence processes take place below ground; their development to the point of surface deformation may involve extended periods of time, and for at least some mechanisms, important indications may lie outside the area directly beneath the surface subsidence. Furthermore, at some sites, the mechanisms at work beneath the surface might be compounded because more than one or two conditions favorable to subsidence occurrence may be present and must be considered in understanding the occurrence and devising corrective or remedial actions.

DID YOU KNOW?

The term "subsidence" is used in this discussion in a broad sense to include both gentle down-warping and the collapse of discrete segments of the ground surface. Keep in mind that displacement is principally downward, although the associated small horizontal components have significant damaging effects. Finally, the term is not restricted on the basis of size of area affected, rate of displacement, or causal mechanism (Allen, 1980).

Land subsidence is a known complement of a variety of natural events that constitute the geologic history of many areas. For pragmatic reasons, geologic processes that are accompanied by subsidence have been examined for evidence that the range in their rates of progress extends into a period that may produce damaging effects in terms of the human time scale. Within this discussion the processes addressed are those that remove or rearrange subsurface materials to produce void space or significant volume reduction—solution, underground erosion, later flow, and compaction—or, in the case of tectonic activity, deep-seated downward displacement. Allen (1980) points out that for all of these naturally occurring geologic processes, examples of related surface subsidence have been found, though some are rare. Where some of these geologic processes are human driven, the incidence of subsidence is greater especially where excavation, loading, or changes in the groundwater regime occur.

Note that for the information provided in this chapter, the basic focus is background information for the single largest cause of subsidence: excessive groundwater pumping. Subsidence due to mining activities is not discussed herein, but several examples of interaction between mining and natural geologic processes are cited. Also, permafrost regions are not discussed, but it is important to note that global climate change is affecting the permafrost to the point where in numerous locations it is melting and turning solid ground areas into ponds or lakes.

DID YOU KNOW?

There are people who have difficulty distinguishing the difference between a sinkhole and land subsidence. Actually, a sinkhole is just one of many forms of ground collapse, or subsidence. Land subsidence is a gradual settling or sudden sinking of the Earth's surface owing to subsurface movement of earth materials. The principal causes of land subsidence are aquifer-system compaction, drainage of organic soils, underground mining, hydrocompaction, natural compaction, sinkholes, and thawing permafrost. Land subsidence can affect areas that are thousands of square miles in size. A sinkhole is a depression in the ground that has no natural external surface drainage. Basically, this means that when it rains, all of the water stays inside the sinkhole and typically drains into the subsurface. Sinkholes are most common in what geologists call "karst terrain." These are regions where the type of rock below the land surface can naturally be dissolved by groundwater circulating through it. Soluble rocks include salt beds and domes, gypsum, limestone, and other carbonate rock (USGS, 2021).

SOLUTION AND SUBSIDENCE

Ordinary soluble components of earth materials that may be connected with subsidence include salt, gypsum, and the carbonate rocks—limestone and dolomite. Note that the roles that these soluble materials perform in the development of surface subsidence depends in part on the degree of their solubility and in part on other physical characteristics.

SALT

Although rock salt (sodium chloride) is one of the most soluble of the common earth materials, the presence of underlying salt deposits has only rarely been associated with surface subsidence under natural conditions in recent times. In these recent times, subsidence in salt deposits is almost always associated with human activity such as mining.

GYPSUM

Gypsum is a soluble rock-forming mineral, which, with its anhydrous counterpart, anhydrite, occurs abundantly in marine evaporite basin deposits. Surface subsidence caused by dissolution of gypsum has occurred in past geologic times. Also, sinkholes have formed in present-day land surfaces underlain by gypsum in New Mexico and Oklahoma. In addition, land subsidence has occurred in areas underlain with rocks and soils with minor amounts of gypsum.

CARBONATE ROCKS

The most widespread incidence of subsidence occurs in carbonate rocks, limestone, and dolomites, not due to their high degree of solubility but because of wide geographic distribution.

SUBSURFACE MECHANICAL EROSION IN SUBSIDENCE

An infrequently recognized phenomenon known as subsurface mechanical erosion (a.k.a. piping) refers to the temporary subsurface flow channels that develop in unconsolidated or friable materials and may lead to surface collapse. Water percolating through pervious surficial materials becomes diverted to a more or less horizontal path to reach the water table or a less pervious stratum. Subsurface water flows whenever it can find its path of least resistance. Water has the uncanny ability to seek out these paths of least resistance.

In the role of subsurface mechanical erosion in subsidence, note that the water, which transports grains of silt and sand, finds an outlet along a nearby valley wall or cliff face or internally in caves, mine openings, or boreholes. Allen (1980) points out that erosion tends to work headward from the outlet, creating and enlarging a tunnel that intersects the vertical flow channel of concentrated percolation water. As tunnel enlargement (as the "pipe" swells) and upward propagation of the roof reduce the support capacity of the surface materials, the ground surface collapses to produce sinkholes.

Three conditions are necessary to produce surface subsidence via the subsurface erosion mechanism: (1) a pervious, easily erodible material must be overlain by material sufficiently competent, at least temporarily, to form a roof above the developing tunnel (the "pipe"); (2) water must have access to the erodible material with sufficient head (i.e., *head*—the vertical distance or height of water above a reference point. Head is usually expressed in feet. In the case of water, head and pressure are related—(1 psi = 2.31 ft of water = pressure) to transport grains of silt and sand; and (3) some sort of outlet must be available for disposal of the flowing water and the sediment grains that it transports (Allen, 1980).

The material that forms the roof of tunnels (of the "pipes") at some localities is a different, and more competent, material than that in the eroded horizon. In other localities, the material forming the roof and the eroded horizon are the same (i.e., loess, altered volcanics), but the competency of the roof is dependent on cohesion in a dry condition provide montmorillonitic clay bonding (i.e., montmorillonitic clay is a very soft phyllosilicate group of minerals that form when they precipitate from water solution as clay). When wet, saturated wet, such cohesion is lost as the component particles become disaggregated.

Although we know about underground mechanical erosion and its impact, we are not adept at the present time to identify cases of it until after the fact. We have found that at least part of the process must be inferred due to the lack of direct observation. Tunnel or pipe development is concealed underground and may only be disclosed by the apparently sudden collapse of surface materials. Unfortunately, the surface collapse is the last step in a long-continued process in which sediments are eroded pebble by pebble, sediment by sediment, and grain by grain and transported to an outlet. Accumulations of transported sediments are rarely observed because the sand and silt grains either become incorporated in the colluvium (i.e., material that accumulates at bottom of steep slope) below the outlet on a valley wall or are washed down into cavities or excavations in the bedrock.

LATERAL FLOW AND SUBSIDENCE

It is uncommon for lateral flow of subsurface materials to cause subsidence, but it is not unknown. Allen (1980) points out that examples have been reported both under natural geologic conditions and under loading by human activities. It's all about plastic flow; those common earth materials susceptible to plastic flow are salt, gypsum, clay, and clay shale.

Geologic examples of subsidence by salt flowage are rim synclines (i.e., a local depression that develops as a perimeter around a salt dome) surrounding salt domes in coastal Texas and Louisiana (Anderson et al., 1973) and broad synclines associated with salt tectonics in the Paradox Basin in Utah and Colorado (Hite and Lohman, 1973). Where the Green and Colorado Rivers have cut deep canyons well down into the formation overlying salt and gypsum in Utah, the removal of load has resulted in local folds and grabens formed when the salt and gypsum flowed laterally (Baker, 1933).

A subsidence feature in east-central England has formed via cambering (i.e., response to stress relief or unloading that results from rapid incision or erosion of the landscape in conjunction with gravitational forces; Hutchinson, 1991) in the Jurassic iron ore.

In the Great Lakes region of the United States, the thick glacial clay deposits have been induced into lateral flows beneath stockpiles of ore, resulting in slight lowering of the ground surface and increasing the distance between ore-retaining walls over a few decades by nearly 2 m (Terzaghi and Peck, 1967).

SUBSIDENCE BY COMPACTION

According to the American Geological Institute (1957, p. 58) compaction is a "decrease in volume of sediments, as a result of compressive stress, usually resulting from continued deposition above them." In this book, *compaction* is defined as the decrease in thickness of sediments, as a result of increase in vertical compressive stress and is synonymous with "one-dimension consolidation" as used by engineers. Note that the term "compaction" is applied both to the process and to the measured change in thickness.

Compaction of sediments in response to increase in applied stress is *elastic* if the applied stress increase is in the stress range less than preconsolidation stress and is *virgin* if the applied stress increase is in the stress range greater than preconsolidation stress.

Elastic compaction (or expansion) is approximately proportional to the change in effective stress over a moderate range of stress and is fully recoverable if the stress reverts to the initial condition. Elastic changes occur almost instantaneously in permeable sediments and, for stresses less than preconsolidation stress, with a relatively small-time delay in strata of low permeability.

Virgin compaction has two components: an inelastic component that is not recoverable upon decrease in stress and a recoverable elastic component. Virgin compaction of aquitards is usually roughly proportional to the logarithm of effective stress increase. In aquitards (fine-grained beds), virgin compaction in response to a humanmade increase in applied stress beyond the preconsolidation stress is a delayed process involving the slow expulsion of pore water and the gradual conversion of the increased applied stress to an increased effective stress. Until sufficient time has passed for excess pore pressure to decrease to zero, measured values of compaction are less than ultimate values. In virgin compaction of aquitards, the inelastic component commonly is many times larger than the elastic component. In coarse-grained beds, on the other hand, the inelastic component may be small compared to the elastic component.

Residual compaction is compaction that would occur ultimately if a given increase in applied stress were maintained until steady-state pore pressures were achieved but had not occurred as of a specified time because excess pore pressures still existed in beds of low diffusivity in the compacting system. It also can be defined as the difference between (1) the amount of competition that will occur ultimately for a given increase in applied stress and (2) that which has occurred at a specified time.

Specific compaction is the decrease in thickness of deposits, per unit of increase in applied stress, during a specified time period.

Specific unit of compaction is the compaction of deposits, per unit of thickness, per unit of increase in applied stress, during a specified time period. Ultimate specific unit compaction is attained when pore pressures in the aquitards have reached hydraulic equilibrium with pore pressures in contiguous aquifers; at that time, specific unit compaction equals gross compressibility of the system.

The *compaction unit* is the compaction per unit thickness of the compacting deposits, usually computed as the measured compaction in a given depth interval during a period of time, divided by the thickness of the interval.

Reduction in the volume of low-density sedimentary deposits that accompanies the process of compaction is a common occurrence, in which particles become more closely packed and the amount of pore space is reduced. The pressures or stresses causing compaction are usually expressed in equivalent "feet of water head" (1 foot of water = 0.433 psi [pounds per square inch]). Compaction may be induced by several actions: by loading, by drainage, by vibration, by extraction of pore fluids, and under certain conditions by the application of water via flooding or other sources. Note that compaction occurs both naturally and by human manipulation.

It's all about pore space. Moreover, the amount of subsidence effected by compaction is a function of the relative amount of pore space in the material as originally deposited, the effectiveness of the compacting mechanism, and the thickness of the deposit undergoing compaction. Natural deposits of unusually high initial porosity include modern delta deposits, terrigenous mudflows, undisturbed loess, and peat.

With regard to *loading* and compaction, the effects of natural loading are most apparent where great thicknesses of fine-grained sediments mount up in a hurry. The process of compaction is accompanied by contemporary subsidence. Currently, on the Mississippi delta, up to 500 million tons of sediment is deposited each year. Fisk et al. (1954) found that levee deposits on the lower delta had subsided 6 meters (19.7 ft) and interdistributary marsh deposits 8.5 meters (27.9 feet).

Drainage is another compaction factor. In low-lying areas, for example, lowering of the water table by artificial drainage stimulates compaction of sediments with accompanying subsidence of the surface. When drainage of organics takes place, local subsidence invariably occurs when soils rich in organic carbon—organic soils—are drained for agriculture or other purposes. Note that the most compelling cause of this subsidence is microbial decomposition which, under drained conditions, readily converts organic carbon to carbon-dioxide gas and water. In addition, compaction, desiccation, erosion by wind and water, and prescribed or accidental burning can also be significant factors.

Let's focus a bit more on organic soils. In the United States of soil taxonomy, organic soils or histosols are one of ten soil orders.

So how much of the United States area consists of organic soils? The total area of organic soils in the United States is roughly equivalent to the size of Minnesota, about 80,000 square miles, nearly half of which is "moss peat" located in Alaska (Lucas, 1982). In the contiguous 48 states about 70 percent of the organic-soil area occurs in northerly, formerly glaciated areas, where moss peats are also common (Stephens et al., 1984). Note that moss peat is composed mainly of sphagnum moss and associated species. It is generally very acidic (pH 3.4 to 4) and, therefore, not readily decomposed, even when drained. However, when amended for agricultural cultivation (for example, through fertilization and heavy application of lime to raise the pH), it can decompose nearly as rapidly as other types of organic soils. This is problematic because in areas such as the Sacramento-San Joaquin Delta of California and the Florida Everglades, containing organic-soil subsidence threatens agricultural production, affects engineering infrastructure that transfers water supplies to large urban populations, and complicates ongoing ecosystem-restoration efforts sponsored by state and federal governments (Galloway et al., 1999).

In low-lying areas, lowering of the water table by artificial drainage may stimulate compaction with sediments with accompanying subsidence of the surface. This is a serious matter, as demonstrated in various areas of the globe, and this is especially the case in the polders of the Netherlands (i.e., in areas where land has been reclaimed from the sea by using dikes and pumps). Bennema et al. (1954) found that clay deposits containing 30 to 35 percent of minus 2-micrometer fraction compressed to about half their original thickness after reclamation over a 100-year period. Note that sediments with about 20 percent fine fraction compressed about 25 percent; compaction of sand layers was negligible.

Drainage of peat areas can commonly result in subsidence for two reasons. First, peat is commonly underlain by, and frequently interbedded with, fine sediments that are susceptible to compaction when drained. Second, peat has certain physical and chemical characteristics that lead to extreme volume changes upon drying (Highway Research Board, 1954). The peat soils' volume changes are the results of the drying and wetting processes. The shrinkage of these soils can be observed during the drying process. In natural conditions, vertical shrinkage causes subsidence of the soil surface, whereas horizontal shrinkage causes soil cracks. The rate of peat shrinkage depends on a number of factors, such as rate of decomposition, values of bulk density and ash content for a particular peat type (Ilnicki, 1967). Peat has a water-holding capacity ranging from 300 to 3,000 percent.

Note that the bulk density of peat is extremely low—about 960 kg/m^3 when wet and 64 kg/m^3 when dry. Also low is specific gravity—between 1.0 and 2.0. Moreover, peat undergoes irreversible biochemical changes on drying that reduce volume. In the United States the largest peat areas that have been subsiding following reclamation for agricultural development are the Florida Everglades (Stephens and Speir, 1969) and the delta area at the confluence of the Sacramento and San Joaquin Rivers in California.

Another factor involved with land compaction is *vibration*; this is especially the case in sedimentary soils under natural conditions during earthquakes. Buildings on saturated alluvium or uncompacted fill may subside or settle differentially in reaction to earthquake vibrations. If the foundations are tied to a lower stable stratum, the buildings may appear to rise as the surrounding sediments subside by compaction.

Terzaghi and Peck (1967) cited a variety of sources of vibration by human sources as having produced subsidence by compaction of underlying earth materials. These sources of vibration include heavy equipment operation, heavy rock-crushing equipment, truck traffic, pile driving, and blasting operations.

Hydrocompaction (i.e., subsidence of land due to the application of water) occurs when loose, dry, low-density deposits compact when they are wetted. This process has produced widespread subsidence in extensive areas in North America, Europe, and Asia. Hydrocompaction may occur under natural overburden load or only with the addition of a surcharge load.

Two general types of subsidence of deposits by hydrocompaction are: (1) loose, moisture-deficient alluvial deposits and (2) moisture-deficient loess and related eolian deposits. Such deposits occur in regions where seasonal rainfall seldom, if ever, is sufficient to penetrate below the root zone; as a result, they have remained moisture deficient throughout their post-depositional history and are readily susceptible to hydrocompaction when they are artificially wetted.

Hydrocompaction-driven subsidence is of critical concern in the design and maintenance of aqueducts, buildings, pipelines, highways, and other major engineering structures. Mitigation techniques to minimize damage can be effected by precompacting the deposits before construction.

To this point we have discussed a few of the causal factors involved in land subsidence, but there are two additional major causes of land subsidence that have not been discussed to this point: land subsidence due to extraction of groundwater (i.e., extraction of soil pore fluids) and land fall or subsidence due to tectonic (earthquake) activities. These two elements involved with land subsidence are discussed and described in Chapters 10 and 11 in this book. For now, the bottom line to this point is that subsidence mechanisms are dynamic, ongoing, and continuing and being analyzed on a continuing basis.

NOTE

1. Based on material in F. Spellman's (2020) *The Science of Environmental Pollution*, 3rd ed. Boca Raton, FL: CRC Press.

REFERENCES

Allen, A.S. (1980). Types of land subsidence. Accessed 02/18/2022 @ www.rcami.lor.usgs. gov/unesco/pdf chapter/chapter8.pdf.
American Geological Institute. (1957). *Glossary of Geology and Related Sciences.* Washington, DC: Am. Geol. Inst., Natl. Acad. Sci-Natl. Acad. Sci-Natl. Research Council, Pub., 591, 325 p.
Anderson, R.E., Eargle, D.H., and Davis, B.O. (1973). *Geologic and Hydrologic Summary of Salt Domes in Gulf Coast Region of Texas, Louisiana, Mississippi, and Alabama.* Denver, CO: USGS.
Baker, A.A. (1933). Geology and oil possibilities of the Moab district, Grand and San Juan Counties, Utah. *U.S. Geol. Survey Bull.* 841:95 p.
Bennema, J., Geuze, E.C.W.A., Skits, H., and Wiggers, A.J. (1954). Soil compaction in relation to quaternary movements of sea-level and subsidence of the land especially in the Netherlands. *Geologie on Mijnbow*, new ser. 16(6):173–178.
Fisk, H.N., McFarlan, E., Jr., Kolb, C.R., and Wilbert, L.J. (1954). Sedimentary framework of the modern Mississippi delta. *Jour. Sed. Petrology* 24(2):76–99.
Franck, I., and Brownstone, D. (1992). *The Green Encyclopedia.* New York: Prentice-Hall.
Galloway, D.I., Jones, D.R., and Ingebritsen, S.E. (eds.) (1999). Land subsidence in the United States: U.S. Geological Survey Circular 1182. 177 p. Accessed @ http://pubs.usgs.gov/circ/circ1182/.
Highway Research Board. (1954). Survey and treatment of marsh deposits. Natl. Research Council. Highway Research Board Bibliography 15, Pub. 314, 95 p.
Hite, R.J., and Lohman, S.W. (1973). *Geologic Appraisal of Paradox Basin Salt Deposits for Waste Emplacement.* Denver, CO: USGS.
Hutchinson, J.N. (1991). Theme lecture: Periglacial and slope processes. *Geologic Society, London, Engineering Geology Special Publications*, 7:283–331.
Ilnicki, P. (1967). The shrinkage of peat soils drying as dependent on soil structure and physical properties. *Zesz. Prob. Post. Nauk. Rol.* 76:197–311 (in Polish).
Kemmer, F.N. (1979). *NALCO Water Handbook.* Fair Oaks, CA: Abe books.

Lucas, R.E. (1982). Organic soils (Histosols): Formation, distribution, physical and chemical properties, and management for crop production: Michigan State University Farm Science Research Report 435, 77 p.

Spellman, F.R., and Whiting, N. (2006). *Environmental Science and Technology: Concepts and Applications.* Boca Raton, FL: CRC Press.

Stephens, J.C., Allen, I.H., Jr., and Chen, E. (1984). Organic soil subsidence. In Holzer, T.L., editor. *Man-Induced Land Subsidence: Geological Society of America Reviews in Engineering Geology*, Vol. 6. Boulder, CO: Geological Society of America, pp. 107–122.

Stephens, J.C., and Speir, W.H. (1969). Subsidence of organic soils in the U.S.A. In Tison, L.J., editor. *Land Subsidence*, Vol. 1. Internat. Assoc. Sci. Hydrology, Pb. No. 89. New York: International Association Science Hydrology, pp. 523–534.

Terzaghi, K., and Peck, R.B. (1967). *Soil Mechanics in Engineering Practices* (2nd ed.). New York: John Wiley, 729 p.

USGS. (2021). *What Is the Difference Between a Sinkhole and Land Subsidence?* Washington, DC: United States Geological Survey.

SUGGESTED READINGS

Terzaghi, K. (1925). Principles of soil mechanics, IV—Settlement and consolidation of clay. *Engineering New-Record*, 95(3):874–878.

10 Relative Sea Level Rise

The Sunday, February 20, 2022, headline in *The Virginian-Pilot*'s editorial opinion page was:

A WORRISOME, WATERY FUTURE

New Estimates of Sea-Level Rise Should Sharpen Focus on the Crisis Facing Hampton Roads

This opinion piece refers to the relative sea level rise occurring in this southeastern Virginia location in the lower Chesapeake Bay Region with the Elizabeth, James, and York Rivers outfalling into the Bay. The article points out that if anyone wants to see the future of Hampton Roads (a.k.a. Tidewater area, composed of the cities Virginia Beach, Norfolk, Chesapeake, Suffolk, Portsmouth, Newport News, Hampton, Williamsburg, and Poquoson), they should refer to coast.noaa.gov (2022), which states that

> $106 billion worth of coastal property will likely be below sea level by 2050 (if we continue on the current path); $12 billion per year will be the cost to ratepayers for new power generation; crop losses could exceed 20 percent over the next 25 years; and, by the end of this century more than $1 trillion worth of coastal property will be below mean sea level or at risk of it during high tide.

These projections, of course, brought about climate via natural or human activities, mean that the migration of humans, land, and aquatic wildlife, with various forms of plant life, will be displaced, forced to migrate to those greener pastures talked about in this book. Unless the millions of inhabitants adopt living in houseboats or another type of floating structure, displacement is likely.

KEY TERMS

Fluid removal from the subsurface leaves voids in the soil under various conditions, and although this chapter focuses on groundwater withdrawal and the subsequent subsidence that it brings about, it is important to point out or to remember that fluid withdrawal also pertains to oil and gas withdrawal.

Anyway, in order to grasp the technical aspects of groundwater or fluid withdrawal and its effect on subsidence (and to therefore understand those aspects of this element of subsidence which lead to displacement and migration of living organisms), it is important to provide key, pertinent terms and definitions right up front in this chapter.

This chapter focuses on an aquifer system that has compacted sufficiently to produce significant and ongoing subsidence in Hampton Roads, Virginia, due to

DOI: 10.1201/9781003295211-10

unconsolidated and semi-consolidated clastic sediments. The definitions given herein are directed toward the type of sediments and breccia within the three-layered (Upper, Middle, Lower) Potomac Aquifer of the Chesapeake Bay Region.

KEY TERMS AND DEFINITIONS[1]

Aquiclude—is an areally (i.e., relating to an area) extensive body of saturated but relatively impermeable material that does not yield appreciable quantities of water to wells. Aquicludes are characterized by very low values of "leakance" (the ratio of vertical hydraulic conductivity to thickness), so that they transmit only minor inter-aquifer flow and also have very low rates of yield from compressible storage. Therefore, they constitute boundaries of aquifer flow systems.

Aquifer system—is a heterogenous body of intercalated (i.e., between) permeable and poorly permeable material that functions regionally as a water-yielding hydraulic unit; it comprises two or more permeable beds separated at least locally by aquitards that impede groundwater movement but don't greatly affect the hydraulic continuity of the system.

Aquitard—is a saturated but poorly permeable bed that impedes groundwater movement and does not yield water freely to wells but may transmit appreciable water to or from adjacent aquifers and, where sufficiently thick, may constitute an important groundwater storage unit. Aquitards are characterized by values of leakance that may range from relatively low to relatively high. Areally extensive aquitards of relatively low leakance may function regionally as boundaries of aquifer flow systems.

Coefficient of volume compressibility (m_v)—is the compression of a lithologic unit, per unit of thickness, per unit increase of effective stress, in the load range exceeding preconsolidation stress (after Terzaghi and Peck, 1948, p. 64).

Consolidation—in soil mechanics, consolidation is the adjustment of a saturated soil in response to increased load, involving the squeezing of water from the pores and a decrease in void ratio (American Society of Civil Engineers, 1962). Note that in this book, the geologic term "compaction" is used in preference to "consolidation," except to report and discuss results of laboratory consolidation tests made in accordance with soil-mechanics techniques.

Excess pore pressure—is transient pore pressure at any point in an aquitard to aquiclude in excess of the pressure that would exist at that point if steady-flow conditions had been attained throughout the bed.

Expansion—the increase in thickness of deposits, per unit of decrease in applied stress. *Specific expansion* is a net specific expansion if compaction is continuing in parts of the interval being measured. The *specific unit of expansion* is the expansion of deposits, per unit of thickness, per unit decrease in applied stress. Specific unit expansion is a net value if compaction is occurring in parts of the interval being measured during the period of decrease in applied stress.

Hydraulic diffusivity—is the ratio of the hydraulic conductivity, K, of a porous medium to unit water-storage capacity (specific storage), S_s, namely K/S_s. The specific storage, S_s, may be defined as the volume of water released from a unit volume of a saturated medium as the result of a unit decline in head. Within the regions of the aquifer system that remain saturated, S_s comprises two principal components, the expansion of the pore water as head is reduced and the decline in head if the position of the overlying water table remains unchanged. Under these conditions, it may be shown that

$$S_s = \gamma_w \beta_w \eta + \gamma_w \beta_t$$

and

$$\frac{K}{S_s} = \frac{K}{\gamma_w \left(\beta_w \eta + \beta_t \right)},$$

where γ_w is the unit weight of water, β_w is the compressibility of water (reciprocal of the bulk modulus of elasticity), η is the porosity, and β_t is the compressibility of the skeletal structure of the medium for stress changes in the elastic range of response.

In highly compressible fine-grained sediments subjected to stresses exceeding the preconsolidation stress, the component due to compressibility of water becomes relatively insignificant; therefore, in the terminology of soil mechanics, the diffusivity is

$$\frac{K}{\gamma_w m_v} = c_v$$

where c_v is termed the *coefficient of consolidation*, and m_v is the coefficient of volume compressibility of the fine-grained sediment.

At any given point within a s saturated porous medium, the rate at which the head changes in response to a change in the head imposed at some other fixed point in the medium is a function of the hydraulic diffusivity. Thus, the hydraulic diffusivity determines the rate at which a head change of specified magnitude migrates through a porous medium.

Stress—the downward stress imposed at an aquifer boundary is *applied stress*. At any given boundary, the applied stress is the weight (per unit area) of sediments and moisture above the water table, plus the submerged weight (per unit area) of the saturated sediments overlying the boundary, plus or minus the net seepage stress (hydrodynamic drag) generated by downward or upward components, respectively, of flow within the specified saturated sediments.

Note that applied stress differs from effective stress in that it defines only the external stress tending to compact a deposit rather than the grain-to-grain stress at any depth within a compacting deposit. Quantitatively, the stress applied to the top of a saturated stratum differs from the effective stress at any depth within the stratum by the submerged weight (per unit area) of the intervening sediments, plus or minus the seepage stress due to vertical flow within the intervening sediments.

Humanmade *changes in applied stress* are of greater practical significance than the absolute value of applied stress, inasmuch as the sediments, before disturbance, are in a state of strength equilibrium with preexisting natural stresses. Change in applied stress within an aquifer system results from either a change in load at the land surface, a change in the position(s) of the potentiometric surface(s) (confined or unconfined), or both. Change in applied stress is uniform throughout a death interval in the potentiometric surfaces and may be expressed as

$$\Delta p = -(\Delta h_c - \Delta h_u Y_s)$$

where p_a is the applied stress expressed in feet of water, h is the head (assume uniform) in the confined aquifer system, h_u is the head in the overlying unconfined aquifer, and Y_s is the average specific yield (expressed as a decimal fraction) in the interval of water-table fluctuation. Change in stress applied to a fine-grained bed becomes effective in changing the thickness of the bed only as rapidly as the diffusivity of the medium permits decrease of excess pore pressures and thus allows the internal grain-to-grain stress (effective stress) to change.

Effective stress is stress (pressure) that is borne by and transmitted through the grain-to-grain of a deposit and thus affects its porosity or void ratio and other physical properties. In one-dimensional compression, effective stress is the average grain-to-grain load per unit area in a plane normal to the applied stress. At any given depth, the effective stress is the weight (per unit area) of sediments and moisture above the water table, plus the submerged weight (per unit area) of sediments between the water table and the specified depth, plus or minus the seepage stress (hydrodynamic drag) produced by downward or upward components, respectively, of water movement through the saturated sediments above the specified depth. Thus, effective stress may be defined as the algebraic sum of the two body stresses, gravitational stress and seepage stress. Effective stress may also be defined as the difference between geostatic and neutral stress.

In an aquifer system, a given change in applied stress results in an immediate equivalent change in effective stress within the aquifers (coarse-grained beds). The increase in stress applied to an interbedded aquitard, however, becomes an increased effective stress within the aquitard only as rapidly as excess pore pressures can decrease. For the change in applied stress to become fully effective, it may take months or years to reach equilibrium because of the low diffusivity of the aquitards.

Geostatic stress is the total load per unit area of sediments and water above some plane of reference.

Gravitational stress is the downward stress within a body of sediments produced by the weight per unit area of sediments and moisture above the water table plus the submerged (buoyed up) weight per unit area of sediments below the water table. Gravitational stress differs from *geostatic (total) stress* in that, below the water table, it includes only the submerged weight of the deposits, whereas the geostatic stress includes the full weight of the saturated deposits (solids plus contained water).

Neutral stress is fluid pressure exerted equally in all directions at a point in a saturated deposit by the head of water. The neutral stress (pressure) is equal to the pressure head multiplied by the unit weight of water, or

$$Y_w = \gamma_w \cdot h_p$$

where u_w is the neutral pressure, γ_w is the unit weight of water, and h_p is the pressure head (Terzaghi and Peck, 1948, p. 2). Neutral pressure is transmitted to the base of the deposit through the pore water and does not have a measurable influence on the void ratio or on any other mechanical property of the deposits.

The total load per unit area (geostatic stress), p, normal to any horizontal plane of reference in a saturated deposit, comprises two components, a neutral stress, u_w, and an effective stress, p'. Therefore, $p = p' + u_w$.

Preconsolidation stress is the maximum antecedent stress to which a deposit has been subjected and which it can withstand without undergoing additional permanent deformation. Stress changes in the range less than the preconsolidation stress produce elastic deformations of small magnitude. In fine-grained materials, stress increases beyond the preconsolidation stress (i.e., the maximum vertical effective stress the soil has ever experienced) produce much larger deformations that are principally inelastic (nonrecoverable).

Stress seepage is when water flows through a porous medium; force is transferred from the water to the medium by viscous friction. The force transferred to the medium is equal to the loss of hydraulic head. The seepage force is directed in the direction of flow.

The vertical seepage force, F, at the base of a stratum across which a hydraulic head differential can be expressed as:

$$F = (h_t - h_b)\gamma_w \cdot A,$$

where h_t and h_b are the heads at the top and bottom respectively, of the stratum; γ_w is the unit weight; and A is the cross-sectional area normal to the direction of seepage.

Under conditions of steady vertical flow, the seepage force is distributed through the body of the medium in the same way as a gravitational force. The average vertical seepage force per unit volume, J, analogous to average unit weight, is

$$\underline{J} = \frac{F}{A \cdot m} = \frac{(h_t - h_b)\gamma_w}{m}.$$

where m is the thickness of the stratum.

The seepage force per unit area, referred to in this report as the seepage stress, J, is

$$J = \mathrm{J} \cdot m = (h_t - h_b)\gamma$$

This vertical seepage stress is algebraically additive with the gravitational stress at the base of the stratum in question, and the sum is transmitted downward through the granular structure of the aquifer system. If the seepage stress, or pressure, is expressed as an equivalent head of water, then γ_w is not required, and the expression is simply

$$J = h_t = h_b$$

Subsidence—is the sinking or settlement of the land surface due to any of several processes. As commonly used, the term relates to the vertical downward movement of natural surfaces, although small-scale horizontal components may be present. The term does not include landslides, which have large-scale horizontal displacements, or settlement of artificial fills.

Subsidence/head-decline ratio is the ratio between land subsidence and the hydraulic head decline in the coarse-grained beds of the compacting aquifer system.

Unit compaction/head-decline ratio is the ratio between the compaction per unit thickness of the compacting deposits and the head decline in the coarse-grained beds of the compacting aquifer system; it equals specific unit compaction if the observed head decline is a direct measure of increase in applied stress.

GROUNDWATER WITHDRAWAL AND SUBSIDENCE

As mentioned, permanent subsidence can occur when water (and other fluids) stored beneath the Earth's surface is removed by pumping. The reduction of fluid pressure in the pores and cracks of aquifer systems, especially in unconsolidated rocks, is inevitably accompanied by some deformation of the aquifer system. Because the granular structure—the so-called "skeleton"—of the aquifer system is not rigid but more or less compliant, a shift in the balance of support for the overlying material causes the skeleton to deform slightly. Both the aquifers and aquitards that constitute the aquifer system undergo deformation, but to different degrees. It is during the typically slow process of aquitard drainage (when the irreversible compression or consolidation of aquitards occurs) that almost all permanent subsidence takes place (Tolman and Poland, 1940). This concept, known as the aquitard-drainage model, has formed the theoretical basis of successful subsidence investigations.

NOTE TO READERS

Before moving on to a discussion of a current event whereby relative sea level rise is active and the current mitigation procedures being used to combat the rising seas and land subsidence, it is important to review a couple of important parameters; this will set the stage for a real-world example of relative sea level rise that is current and ongoing.

EFFECTIVE STRESS

The principle of effective stress was first proposed by Terzaghi (1925). For our purpose in this book, "effective" means the calculated stress that was effective in moving soil and/or causing displacements. According to this principle, when the support provided by fluid pressure is reduced (such as when groundwater levels are lowered), support previously provided by the pore-fluid pressure is transferred to the skeleton of the aquifer system, which compresses to a degree. On the other hand, when the pore-fluid pressure is increased, such as when groundwater recharges the aquifer system, support previously provided by the skeleton is transferred to the fluid and the skeleton expands. In this way, the skeleton alternatively undergoes compression and expansion as the pore-fluid pressure fluctuates with aquifer-system discharge and recharge. When the load on the skeleton remains less than any previous maximum load, the fluctuations create only a small elastic deformation of the aquifer system and small displacement of land surface. This fully recoverable deformation occurs in all aquifer systems, commonly resulting in seasonal, reversible displacements in land surface of up to 1 inch or more in response to the seasonal changes in groundwater pumpage (USGS, 2013).

PRECONSOLIDATION STRESS

The maximum level of past stressing of a skeletal element is termed the preconsolidation stress. Stated differently, preconsolidation stress is the maximum effective vertical overburden stress that particular soils have sustained in the past. When the load on the aquitard skeleton exceeds the preconsolidation stress, the aquitard skeleton may undergo significant, permanent rearrangement, resulting in irreversible compaction. Because the skeleton defines the pore structure of the aquitards, this results in a permanent reduction of pore volume as the pore fluid is "squeezed" out of the aquitards into the aquifers. In confined aquifer systems subject to large-scale overdraft, the volume or water derived from irreversible aquitard compaction is essentially equal to the volume of subsidence and can typically range from 10 to 30 percent of the total volume of water pumped. This represents a one-time mining of stored groundwater and a small permanent reduction in the storage capacity of the aquifer system. Alternative names of preconsolidation stress are preconsolidation pressure, pre-compression stress, pre-compaction stress, and preload stress (Dawidowski and Koolen, 1994).

AQUITARD ROLE IN COMPACTION

In recent decades, increasing recognition has been given to the critical role of aquitards in the intermediate and long-term response of alluvial systems to ground water pumpage. Aquitard systems play a key role in compaction. In such systems, interbedded layers of silt sand clays (once dismissed as non-water yielding) constitute the bulk of the groundwater storage capacity of the confined aquifer system. This is the case based on their substantially greater porosity and compressibility and, in many cases, their greater aggregate thickness compared to the more transmissive, coarser-grained sand and gravel layers.

Aquitards are less permeable than aquifers. Thus, the vertical drainage of aquitards into adjacent pumped aquifers may proceed very slowly and thus lag far behind the changing water levels in adjacent aquifers. The lagged response within the inner portions of a thick aquitard may be largely isolated from the higher frequency seasonal fluctuations and more influenced by lower-frequency, longer-term trends in groundwater levels. Because the migration of increased internal stress into the aquitard accompanies its drainage, as more fluid is squeezed from the interior of the aquitard, larger and larger internal stresses from the interior of the aquitard propagate farther into the aquitard.

When the preconsolidation stress is exceeded by the internal stresses, the compressibility increases dramatically, typically by a factor of 20 to 100, and the resulting compaction is largely nonrecoverable. At stresses greater than the preconsolidation stress, the lag in aquitard drainage increases by comparable factors, and concomitant compaction may require decades or centuries to approach completion. The theory of hydrodynamic consolidation (Terzaghi, 1925)—an essential element of the aquitard drainage model—describes the delay involved in draining aquitards when heads are lowered in adjacent aquifers, as well as the residual compaction that may continue long after drawdowns in the aquifers have essentially stabilized. Numerical modeling based on Terzaghi's theory has successfully simulated complex histories of compaction observed in response to measure water-level fluctuations (Helm, 1975).

DID YOU KNOW?

Studies of subsidence in the Santa Clara Valley in California established the theoretical and field application of the laboratory-derived principle of effective stress and theory of hydrodynamic consolidation to the drainage and compaction of aquitards (Tolman and Poland, 1940; Poland and Green, 1962; Poland and Ireland, 1988).

DID YOU KNOW?

Responses to changing water levels following decades of groundwater development suggest that stresses directly driving much of the compaction are somewhat insulated from the changing stresses caused by short-term water-level variations in the aquifers.

THE VANISHING OF HAMPTON ROADS[2]

When the name Hampton Roads is mentioned, it is not unusual for unfamiliar people to shake their heads and say, "What?" When and if the name Hampton Roads is roughly familiar to people, they may ask, "Do you mean the body of water . . . or do you mean the location?"

Well, actually, the term Hampton Roads is the name of both a body of water in Virginia and the surrounding metropolitan region in southeastern Virginia and northeastern North Carolina. The land area is also known as Tidewater.

For our purpose in this text, when we refer to Hampton Roads, we are referring to both the water body and land region as one, because land subsidence and relative sea level rise in the area pertain to both. With regard to the total area, it is composed of 527 square miles (1,364 km²) and is made up of nine major cities: Norfolk, Virginia Beach, Chesapeake, Newport News, Hampton, Portsmouth, Suffolk, Poquoson, and Williamsburg; as a combined statistical area, it also includes Kitty Hawk and Elizabeth City, North Carolina. The entire area has a population of over 1.7 million. With regard to the body of water known as Hampton Roads, it is one of the world's largest natural harbors. It incorporates the mouths of the Elizabeth River, Nansemond River, York River, and James River with smaller rivers and streams and empties into the Chesapeake Bay—a treasured estuary—near its mouth leading to the Atlantic Ocean.

NOTE TO READERS

The Chesapeake Bay region is a treasured estuary, as mentioned, but also it is a bastion of early American history. Moreover, it is not a bad place to live; having resided in the area for more than 50 years, I think I am qualified to state this as fact. Having spent those 50 years studying the Bay area and specifically water pollution problems in the area, it has become a lifetime project for me to continue the study, but my focus has shifted dramatically from the pollution of the Bay and the existence of algal dead zones to a more pressing problem: the literal vanishing of the land area and, as a result, the potential for human and wildlife displacement or migration.

LOCATION OF CHESAPEAKE BAY

Thirty-five million years ago the Chesapeake Bay did not exist. Its genesis during this timeframe is the result of a bolide impact that smashed into the area, creating a crater. Even as late as 18,000 years ago, the bay region was dry land; the last great ice sheet was at its maximum over North America, and sea level was about 200 m lower than at present. This sea level exposed the area that now is the bay bottom and continental shelf. Because sea level was so low, the major east coast rivers had to cut narrow valleys across the region all the way to the shelf edge. About 10,000 years ago, however, the ice sheets began to melt rapidly, causing sea level to rise and flood the shelf and the coastal river valleys. The flooded valleys became the major modern estuaries, like Delaware Bay and Chesapeake Bay. To come to the point, the impact crater created a long-lasting topographic depression, which helped determined the eventual location of Chesapeake Bay (USGS, 2016).

RIVER DIVERSION

The rivers of the Chesapeake region converged at a location directly over the buried crater. Some might think the convergence of these rivers is merely coincidence.

Is it?

The short answer: no, it is not coincidence. Notice that in Figure 10.1, the important river channels in the area change course significantly just after they cross the rim of the buried bolide crater. These channels are actually successive buried ice-age channels of the ancient Susquehanna River (formed from 450,000 to 20,000 years ago). This river diversion, along with seismic evidence that post-impact units sagged and thickened over the crater, indicate that the ground surface over the crater remained lower than the areas outside the crater for 35 million years.

The question is, why?

Why does the Rappahannock River flow southeastward to the Atlantic, while the York and James Rivers make sharp turns to the northeast near the outer rim of the crater?

What is the answer?

Well, the courses of the York and James Rivers in the lower Bay region are the result of the ongoing influence of differential subsidence over the bolide crater.

Differential subsidence?

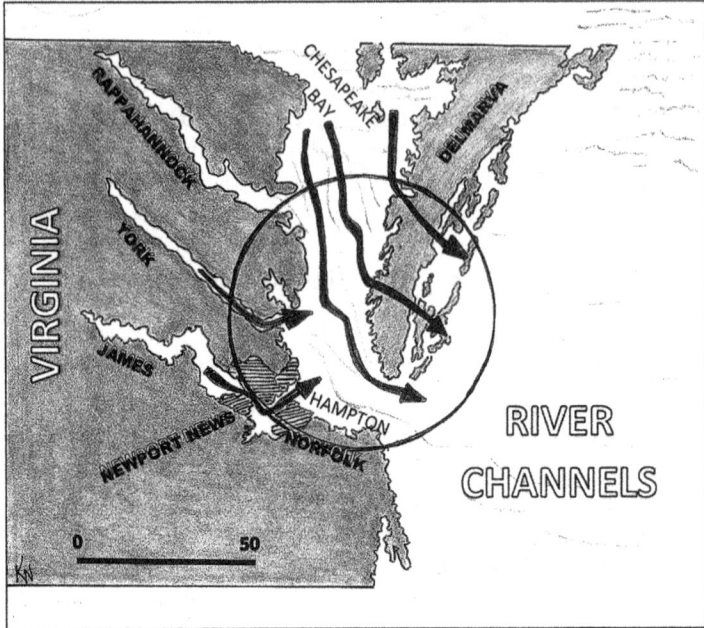

FIGURE 10.1   River channels, Chesapeake Bay region.

Source: Adapted from USGS (1997). Illustration by F. R. Spellman and Kat Welsh.

Yes. Absolutely. Two factors cause subsidence in the region. First, subsidence is the result of loading during the past 35 million years since the impact. Second, subsidence is also due to compaction of the breccia; that is, rock composed of broken fragments of minerals or gravel cemented together by a fine-grained matrix. The crater breccia is 1.2 km thick and was deposited as water-saturated sandy, rubble-bearing, non-jelled jello-like slurry. The sediment layers surrounding it were already partly consolidated, so the mushy breccia would compact much more rapidly under its subsequent sediment load than the surrounding strata.

You may be asking what all this has to do with the local Bay rivers. The combination of the two factors detailed previously produced a subsidence differential, causing the land surface over the breccia to remain lower than the land surface outside the crater. Therefore, the river valleys covered the crater and were located in those particular places when rising sea level flooded them. In short, the impact crater created a long-lasting topographic depression, which helped predetermine the eventual location of Chesapeake Bay (USGS, 1997). Finally, it is important to point out that one of the main focuses of this book and the SWIFT initiative (Sustainable Water Initiative for Tomorrow) is land subsidence in the Chesapeake Bay area; this will be discussed in detail later.

For now, the important point to know and remember is the continued influence of differential subsidence over the crater.

Ground Instability Due to Faulting

Seismic profiles across the crater show faults that cut the sedimentary beds above the breccia and extend upward toward the bay floor (see Figure 10.2). The current resolution of our seismic profiles allows us to trace the faults to within 10 m of the bay floor. These faults represent another result of the differential compaction and subsidence of the breccia. As the breccia continues to subside under the load of post-impact deposits, it subsides unevenly due to its viable content of sand and huge clasts. This eventually causes the overlying beds to bend and break and to slide apart along the fault planes. These faults are zones of crustal weakness and have the potential for continued slow movement or sudden large offsets if reactivated by earthquakes.

Inquiring minds might ask why it is important to know about the faults and their location. It is important for us to know in detail the location, orientation, and amount of offset of these compaction faults because of the potential for the faulting to separate adjacent sides of the confining unit over the salt-water reservoir. If this occurred, it could allow the salty water to flow upward and to contaminate the fresh water supply.

Using the seismic profiles on hand, Dr. Wylie Poag, senior research scientist, U.S. Geological Survey, has identified and is mapping more than 100 faults or fault clusters around and over the crater, which reach to or near the bay floor (USGS, 2016).

Disruption of Coastal Aquifers

The hydrogeological framework thought to be typical of southeastern Virginia, in cross section, consists of groundwater aquifers alternating with confining beds. The

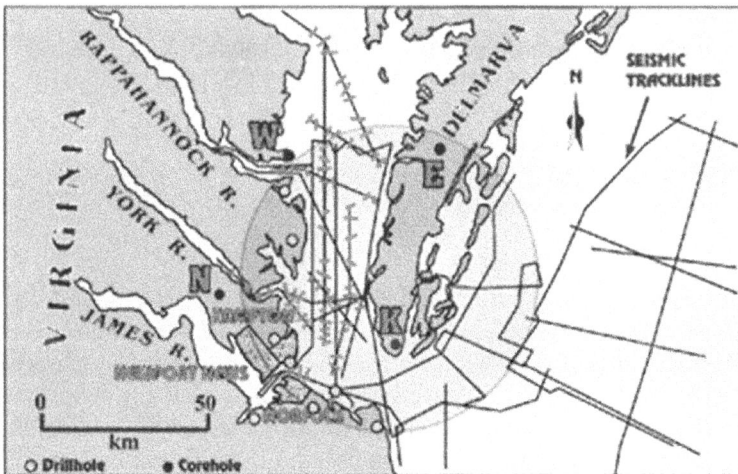

FIGURE 10.2 Image showing the location of faults (dashes) where they cross seismic profiles. The large circle shows the extent of the buried crater. The brick pattern shows the three main cities of the lower Chesapeake Bay. Capital letters mark the locations of Newport News, Windmill Point, Exmore, and Kiptopeke core holes.

Source: USGS (2016).

aquifers are mainly sand beds, which contain water-filled pore spaces between the sand grains. The pore spaces are connected, which allows the water to flow slowly though the aquifers. The confining beds are mainly clay beds, which have only very fine pores. These are poorly interconnected, which greatly retards or prevents the flow of water. Before we knew about the Chesapeake Bay crater, this framework of alternating aquifers and confining units was applied to models of groundwater flow and water-quality assessments in the lower Chesapeake Bay region.

Based on core samples, researchers have determined that in the crater area itself, the orderly stack of aquifers seen outside the crater does not exist; instead, they were truncated and excavated by the bolide impact. In place of those aquifers, there is now a single huge reservoir with a volume of 4,000 km^3. That's enough breccia to cover all of Virginia and Maryland with a layer 30 m thick. But the most startling part is that this huge new reservoir does not contain fresh water like the aquifers it replaced; the pore spaces are filled with briny water that is 1.5 times saltier than normal seawater. This water is too salty to drink or to use in industry (USGS, 2016). It is interesting to note that for decades geohydrologists and others in the Hampton Roads region scratched their collective heads and wondered why, in locations away from the crater, water wells yielded good-quality freshwater suitable for potable purposes; however, whenever wells were drilled within the crater ring or close to it, salty water was all that could be found.

DID YOU KNOW?

The parameters for saline water are:

- Fresh water—Less than 1,000 ppm.
- Slightly saline water—From 1,000 to 3,000 ppm.
- Moderately saline water—From 3,000 to 10,000 ppm.
- Highly saline water—From 10,000 to 35,000 ppm.
- Ocean water—Contains about 35,000 ppm of salt.
- Chesapeake Bay crater water—Contains about 1.5 times more salt than normal seawater.

The presence of this hypersaline aquifer has practical implications for groundwater management in the lower Bay region. For example, we need to know how deeply buried the breccia is in order to avoid drilling into it inadvertently and contaminating the overlying freshwater aquifers. Its presence also limits the availability of freshwater. On the Delmarva Peninsula, over the deepest part of the crater, only the aquifers above the breccia are available for freshwater. The bolide crater investigation shows that we need to be especially conservative of groundwater use in the area (USGS, 2016).

LAND SUBSIDENCE

Land subsidence and the potential for land rebound provided by injecting treated wastewater to drinking water quality into the Potomac Aquifer is the focus of this book. Later, much will be said about this topic and how it is related to the Chesapeake Bay bolide impact crater and its effects, along with relative sea level rise occurring at the present time. For now, it is important to point out that there is growing evidence that accelerated land subsidence is reflected in the geology and topography of the modern land surfaces around the bolide crater. The breccia is 1.3 km thick and was deposited as water-saturated, sandy, rubble-bearing slurry (like concrete before it hardens). The sediment layers surrounding the crater, on the other hand, were already partly consolidated, so the mushy breccia compacted much more rapidly under its subsequent sediment load than the surrounding strata. The compaction differences produce a subsidence differential (i.e., the difference in subsidence between two points on the crater), causing the land surface over the breccia (due to breccia compaction) to remain lower than the land surface over sediments outside the crater.

During Dr. Poag's investigation, he and his team observed that the boundary between older surface rocks and younger surface rocks coincides with the position and orientation of the crater rim on all three peninsulas that cross the rim. The older beds have sagged over the subsiding breccia, and the younger rocks have been deposited in the resulting topographic depression. The topography also reflects the differential subsidence. The Suffolk Scarp and the Ames Ridge are elevated landforms (10 to 15 meters high) located at, and oriented parallel to, the crater rim.

Crater-related ground subsidence also may play a role in the high rate of relative sea-level rise documented for the Chesapeake Bay region. One of the locations of highest relative sea-level rise is Hampton Roads (the lower part of the James River, located over the crater rim).

The preceding information has set the first row of foundation blocks for the material to follow. Specifically, this chapter describes the late Eocene period in the Virginia Coastal Plain area when the formerly quiescent geological regime was dramatically transformed when a bolide struck in the vicinity of the Delmarva Peninsula. This consequential event produced the following principal consequences (USGS, 1998):

- The bolide carved a roughly circular crater twice the size of the state of Rhode Island (~6,400 km^2) and nearly as deep as the Grand Canyon (1.3 km deep).
- The excavation truncated all existing groundwater aquifers in the impact area by gouging ~4,300 km^3 of rock from the upper lithosphere, including Proterozoic and Paleozoic crystalline basement rocks and Middle Jurassic to upper Eocene sedimentary rocks.
- A structural and topographic low formed over the crater.
- The impact crater may have predetermined the present-day location of Chesapeake Bay.
- A porous breccia lens, 600–1,200 m thick, replaced local aquifers, resulting in groundwater ~1.5 times saltier than normal sea water.
- Long-term differential compaction and subsidence of the breccia lens spawned extensive fault systems in the area, which are potential hazards for local population centers in the Chesapeake Bay area.

HAMPTON ROADS: SEA LEVEL RISE

Of all the potential impacts of natural (cyclical) or human-induced climate change, a global rise in sea level appears to be the most certain and the most dramatic. As shown in Figure 10.3, for the last 5,000 years, the rate or sea level rise was only 3 feet per 1,000 years. In the Chesapeake Bay region, the relative rise in sea level has been about 1 foot during the last 100 years (Figure 10.3). While scientists view this rapid rate as possibly a temporary acceleration, these same scientists believe that it signals a new trend in response to global warming. The point is, if the rate of rise accelerates in the near future as projected, it could have serious repercussions for Chesapeake Bay.

Water levels are measured relative to the land, and as stated earlier, relative sea-level rise in the Chesapeake Bay region has two components: global water level increase and land subsidence. Worldwide or eustatic seal-level rise is caused by water released from melting glaciers and thermal expansion of seawater. Both are related to global warming and have amounted to a 6-inch rise in the last century. Let's take a closer look at the global climate change and warming problem.

Land Subsidence Increases Flooding Risk

As relative sea levels rise, shorelines retreat, and the magnitude and frequency of near-shore coastal flooding increases. This is particularly a problem in Norfolk, Virginia (downtown area and Ocean View District), where during a coastal storm event and corresponding high tide, downtown Norfolk streets flood, at times 3–5 feet. Although land subsidence can be slow, its effects accumulate over time; this has been an expensive problem in the Norfolk area and other parts of the southern Chesapeake Bay region. Analysis by McFarlane (2012) found that between 59,000 and 176,000 residents living near the shores of the southern Chesapeake Bay could be either permanently inundated or regularly flooded by 2100. This estimate was

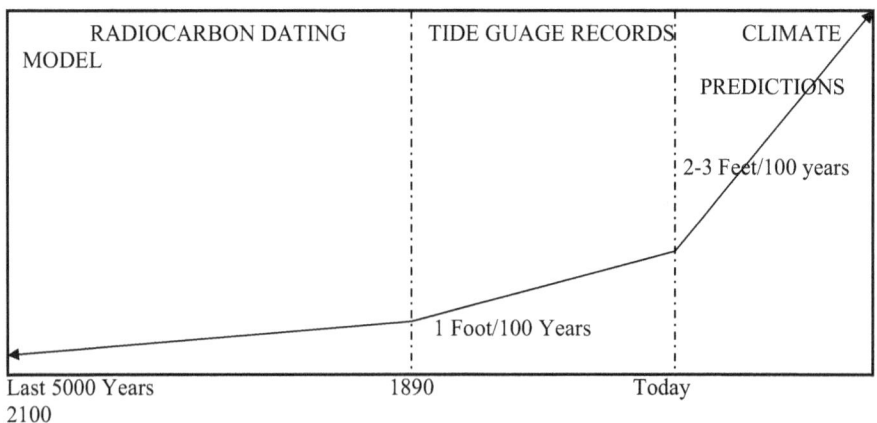

FIGURE 10.3   Sea-level rise in Hampton Roads region past and present.

Source: USFWS (1995).

based on the 2010 census data, using the spring high tide as a reference elevation and assuming a 1-m relative sea-level rise. Damage to private property was estimated to be $9 to $26 billion, and 120,000 acres of economically valuable land could be inundated or regularly flooded under these same assumptions. Historic and cultural resources are also vulnerable to increased flooding from relative sea-level rise in the southern Chesapeake Bay, particularly at shoreline sites near tidal water, such as the 17th-century historic Jamestown site.

It should be pointed out that the shoreline area in southern Hampton Roads is not the only area prone to flooding. Land subsidence can also increase flooding in areas away from the coast in low-lying areas such as Franklin, Virginia. The city of Franklin is about 60 road miles west of Hampton Roads. The Blackwater River Basin, which encompasses Franklin and other local areas, can be subject to increased flooding as the land sinks. In fact, Franklin and the counties of Isle of Wight and Southampton have experienced large floods (Federal Emergency Management Agency, 2002). Land subsidence may be altering the topographic gradient that drives the flow of the river and possibly contributing to the flooding.

WETLAND AND COASTAL MARSH ECOSYSTEMS

Wetland and marsh ecosystems in low-lying coastal areas are sensitive to slight changes in elevation (Cahoon et al., 2009). Salt marshes, which are widespread in the southern Chesapeake Bay region, are dependent on tidal dynamics for their existence. Slight changes in either land or sea elevations can alter sediment deposition, organic production and plant growth, and the balance between fresh water and seawater (Morris et al., 2002). The effects of sea-level rise on tidal wetlands are frequent and already apparent in local wetlands. These effects include:

- Shoreline erosion
- Habitat loss
- Changes in tidal amplitude
- Landward migration of tidal waters
- Landward migration of habitats
- More frequent inundation
- Changes in plant and animal species composition
- Changes in tidal flow patterns
- Migration of estuarine salinity gradients
- Changes in sediment transport

Although sea-level rise has one of the most direct effects on tidal wetlands, shoreline environments also are affected by land subsidence. When land subsides, it subjects shorelines to increased wave action, increasing erosion and wash-over. This type of damage is happening in the Chesapeake Bay because of relative sea-level rise (Erwin et al., 2011); Kirwan and Guntenspergen, 2012; Kirwan et al., 2012). Major changes in the coastal and marine ecosystem of the southern Chesapeake Bay are expected to be caused by relative sea-level rise (Cahoon et al., 2009); these changes will likely be more severe if land subsidence continues.

LAND SUBSIDENCE CAN DAMAGE INFRASTRUCTURE

Buildings, bridges, canals, water and wastewater treatment plants, electrical substations, communication towers, pipes, and other components that make up a region's infrastructure can be damaged from relative groundwater rise or from differential settling in areas with high subsidence gradients (Galloway et al., 1999). As land sinks and sea level continues to rise, groundwater levels rise towards the land surface in coastal areas, which can cause problems for subterranean structures, septic fields, buried pipes and tanks and cables, and infrastructure not designed for elevated groundwater levels. Storm and wastewater interceptor lines in urban areas are vulnerable because land subsidence can alter the topographic gradient driving the flow through the sewers, causing increased flooding and more frequent sewage discharge from combined sewer overflows.

GAUGING SUBSIDENCE[3]

Land subsidence can be effectively and accurately measured using several reliable and proven techniques. Multiple measuring or monitoring techniques are often used together to understand different aspects of land subsidence (Table 10.1). Because rates and locations of land subsidence change over time, repeat measurements at multiple locations are often needed to improve understanding of the complex phenomenon and guide computer models that forecast future subsidence. Extensometers measure changes in aquifer-system thickness, whereas other methods measure land surface elevation, from which subsidence is calculated by subtracting measurements over time.

TABLE 10.1
Land Subsidence Monitoring Methods

Method	Type of Data	Measures Aquifer System Compaction Independently	Spatial Coverage	Temporal Detail
Borehole extensometer	Aquifer-system thickness at one location, continuous record	Yes	Low	High
Tidal station	Sea elevations at one location, continuous record	No	Low	High
Geodetic Surveying	Land elevations at one or several locations, multiple times or continuous record	No	Low to moderate	Low to high
Remote sensing (InSAR)	Land elevations over a wide area at multiple times	No	High	Moderate

Source: USGS (2013).

GPS—global positioning system, InSAR—interferometric synthetic aperture radar.

Borehole Extensometers

An *extensometer* is a device that is used to measure changes in the length of an object. Used in applications related to land subsidence measurement, a borehole extensometer measures compaction or expansion of an aquifer system independently of other vertical movements, such as crustal and tectonic motions (Galloway et al., 1999). An extensometer measures change in aquifer-system thickness by recording changes in the distance between two points in a well (see Figure 10.4). Usually, the

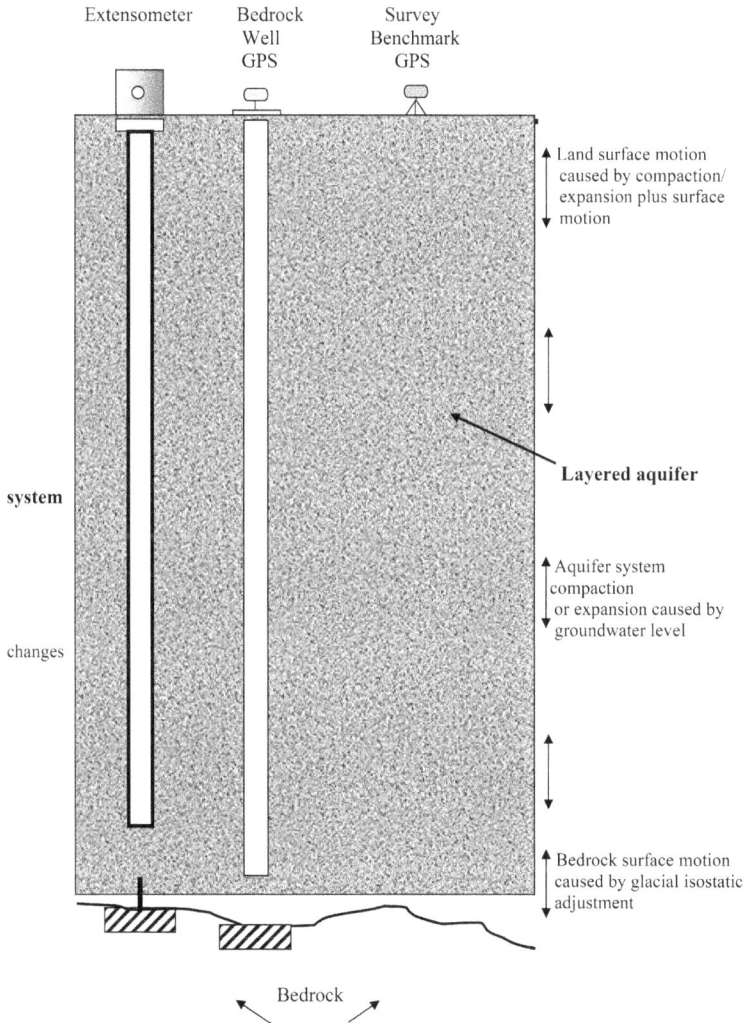

FIGURE 10.4 Subsidence monitoring methods. Survey benchmark GPS to measure land surface motion. Bedrock well GPS to measure bedrock surface motion and extensometer to measure aquifer system compaction or expansion.

Source: USGS (2016).

two measurement points are established at the top and bottom of a well to measure total aquifer-system compaction between the land surface and the bottom of the aquifer system. Alternatively, specific intervals within a well can be measured, for example, to measure compaction of just one aquifer within a layered aquifer system. Extensometer measurements are often combined with surface monitoring techniques to determine the portion of total land subsidence attributable to aquifer-system compaction (Poland, 1984).

HRSD's extensometer used as part of its SWIFT project, installed at Nansemond Treatment Plant in Suffolk, Virginia, and owned, operated, recorded, and maintained by USGS, is shown in Figures 10.5, 10.6, and 10.7.

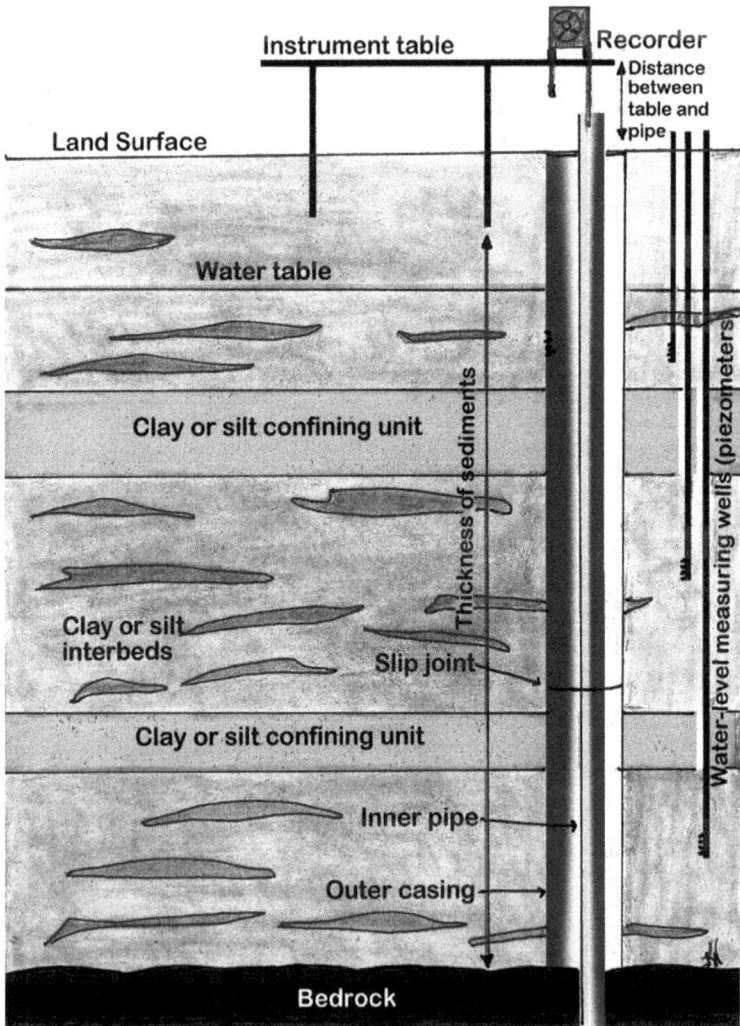

FIGURE 10.5   A borehole extensometer with recording equipment.

Source: USGS (2013). Illustration by F.R. Spellman and K. Welsh.

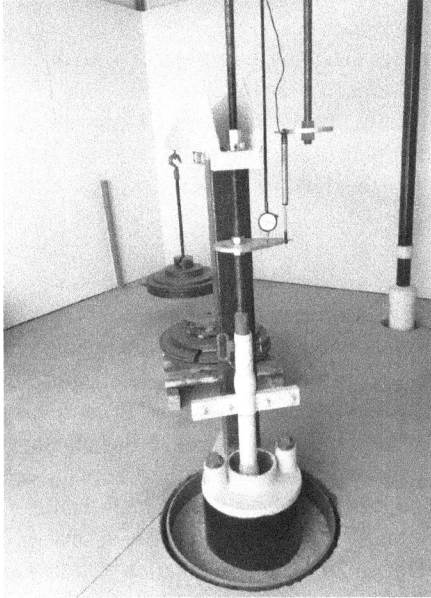

FIGURE 10.6   USGS HRSD Nansemond pipe extensometer with a total depth of 1,960 feet.

Source: Public Domain Photo. By D. L. Nelms Hydrologist USGS (2016). Accessed 08/05/2022 @ www.usgs.gov/media/images/usgs-nansemond-pipe-extensometer.

FIGURE 10.7   The housing of the Nansemond Treatment Plant extensometer showing the triangular table and instrument bridge above the extensometer. The piers that support the table extend down 65 feet. The movement of the table relative to the extensometer is how land-surface movement is measured.

Source: Public domain photo. By D.L. Nelms Hydrologist USGS (2016). Accessed 08/05/2022 @ www.usgs.gov/media/images/usgs-nansemond-pipe-extensometer-0.

TIDAL STATIONS

A tidal station measures sea elevation at one location. To determine long-term trends, sea-level measures are averaged over time to remove the effect of waves, tides, and other short-term fluctuations. In the southern Chesapeake Bay, tidal stations have been in operation for many decades (Table 10.2). Tidal stations are significant and valuable because they indicate relative sea-level rise.

GEODETIC SURVEY

Geodetic surveying is the measurement of land surface coordinates. Geodetic surveying is most commonly performed using either global positioning system (GPS) technology that reads signals from satellites to obtain very detailed location and time information or with traditional optical leveling equipment. Because historical geodetic survey records are available for the southern Chesapeake Bay region, geodetic surveying can be used to determine cumulative land subsidence over many decades. Benchmark stations are established and, for as long as they remain undisturbed, can be surveyed multiple times to determine elevation changes between surveys (USGS, 2013).

The Continuously Operating Reference Station (CORS) network, a network of long-term GPS stations throughout the United States that includes stations in the Southern Chesapeake Bay region, is managed by the National Geodetic Survey. Each CORS station continuously records three-dimensional position data (north–south, east–west, and up–down), allowing rates of change to be calculated over time (Snay and Soler, 2008). A CORS station records ground position at one site and is designed to operate for many years.

Besides stationary GPS sites such as the CORS stations, portable GPS receivers can be used to expand spatial coverage. In the Houston-Galveston area, GPS receivers mounted on trailers have been used to collect data at up to four different sites each month (Galloway et al., 1999; Bawden et al., 2003). The portable GPS approach has acceptable

---

TABLE 10.2

Relative Sea-Level Rise at Selected National Oceanic and Atmospheric Administration Tidal Stations in the Southern Chesapeake Bay Region

| | | | Rate of Relative Sea-Level Rise | |
| | | | Measured, | |
ID	Site Name	Period	(mm/yr)	95% CI
8632200	Kiptopeke, Virginia	1951–2006	3.5	±0.42
8637624	Gloucester Point, Virginia	1950–2006	3.8	±0.47
8638610	Sewells Point, Virginia	1927–2006	4.4	±0.27
8638660	Portsmouth, Virginia	1935–2006	3.8	±0.45
	Average		**3.9**	**±0.40**

Source: Data are from Zervas (2009); USGS (2013). CI, confidence interval; mm, millimeters; mm/yr, millimeters per year; %, percent.

subcentimeter accuracy, gives better spatial coverage than stationary GPS would, and has a lower cost than interferometric synthetic aperture radar (InSAR) technology.

InSAR

Interferometric synthetic aperture radar is a radar technique used in geodesy and remote sensing. InSAR has been used to investigate surface deformation resulting from land subsidence (Galloway and Hoffman, 2007). With InSAR, as little as 5 mm of elevation change can be measured over hundreds or thousands of square kilometers with a horizontal spatial resolution down to 20 m (Pritchard, 2006). Interferograms (maps) show land-surface elevation changes that are produced by combining two synthetic aperture rate (SAR) images acquired by multiple satellite or airborne passes over the same area at different times. InSAR analysis has the advantage of measuring subsidence over a large area, whereas traditional geodetic leveling and GPS surveying are performed at only one or a handful of locations during a survey (Sneed et al., 2002; Stork and Sneed, 2002).

Using InSAR in the Chesapeake Bay region has potential limitations. Subsidence rates determined by InSAR might have errors that are larger than the subsidence rates observed in the region (1.1 to 4.8 mm/yr). The region's high humidity and dense vegetation would create spurious radar signals, require the use of persistent scatter techniques, and result in lower measurement resolution than is found in more arid regions. Also, available satellite data cover only a relatively short time span. The best available synthetic aperture radar satellite data for the southern Chesapeake Bay region cover from 1992 to 2000, so the time of accumulated subsidence determined from these data would be no more than 8 years. Despite these limitations, InSAR could be used to identify hotspot areas of subsidence. Such mapping could be useful for identifying unexpected areas of subsidence, focusing attention on important areas, and picking locations for other ground-based subsidence monitoring techniques (USGS, 2013).

IMPORTANCE OF LAND SUBSIDENCE MONITORING

Land subsidence, or sinking of the land surface, has occurred, and is still occurring, in many locations throughout the United States. Earlier, we pointed out a few of the most impacted areas in the United States, but the focus of this book is on Hampton Roads, Virginia (the area of southern Chesapeake Bay, the largest estuary in the United States), where land subsidence is an ongoing and serious issue but also where innovation and technology and far thinking might be able to arrest or correct the problem. In Hampton Roads, the boundary between land and water is low lying. Land subsidence is important in the region because it causes increased flooding; alters wetland and coastal ecosystems; and damages infrastructure and historical sites such as Jamestown, Virginia.

As population densities continue to increase in the Hampton Roads region, the flood hazard unfortunately is increasing as well, due to locally high rates of land subsidence with global sea level rise. The combination—global sea level rise and coastal subsidence—is doubly significant even in locations where the occurrence of major hurricanes is relatively rare compared to other regions of the country. In the Hampton Roads region in particular, extratropical cyclones or "nor'easters" that have not caused significant flooding in the past will begin to do so—and with greater frequency—as sea level continues to rise *relative* to the land. Flood hazard mitigation compounded by land subsidence and relative sea level rise at a minimum requires an understanding of the land and by-ocean processes that contribute to it in complex and often unpredictable ways.

The fact is that rates and locations of land subsidence are not well known throughout the Hampton Roads area because monitoring has been insufficient in recent decades. Monitoring data are needed to better understand rates and locations of land subsidence and to plan for preventing or mitigating its potentially damaging effects.

Again, recurrent flooding problems have prompted concern about land subsidence in Hampton Roads (Sweet et al., 2014). In addition, these concerns are compounded by evidence that groundwater pumping and associated aquifer depressurization have caused past land subsidence (Pope and Burbey, 2004; Holdahl and Morrison, 1974) and measurements showing that relative sea-level rise is faster in Hampton Roads that elsewhere on the Atlantic coast (Sallenger et al., 2012). As mentioned, rates and locations of land subsidence are not well known through the Hampton Roads area because monitoring has been insufficient. Therefore, risks commonly associated with coastal land subsidence—increased flooding, alteration of wetland and coastal ecosystems, and damage to infrastructure and historical sites—cannot be accurately assessed. More frequent monitoring at multiple locations using multiple complementary methods is needed to build an understanding of subsidence and to plan how to avoid or mitigate the effects of subsidence.

Before land subsidence can be understood, it must be monitored. Monitoring data provide the foundation for understanding why, where, and how fast land subsidence is occurring, both now and in the future. Because rates of land subsidence change over time and vary from one location to another, monitoring should be done at multiple locations for multiple years. Monitoring data are used (USGS, 2016)

- To avoid or mitigate problems caused by land subsidence—Urban planners, resource managers, and politicians use monitoring data to guide their decisions.
- To answer questions—Such as why is subsidence occurring?
- To predict future land subsidence—Predictive models that can test mitigation strategies require monitoring data for accuracy and reliability.
- To make maps—Maps showing critical areas for mitigating land subsidence are based on monitoring data.
- Land subsidence monitoring measures
  - land surface motion
  - bedrock surface motion
  - changes in aquifer system thickness

MONITORING METHODS

Land subsidence is detected by measuring land surface positions over time and calculating rates of change by subtraction. As pointed out earlier, there are several reliable and accurate techniques for measuring land subsidence in Hampton Roads: borehole extensometers, tidal stations, geodetic surveying, and remote sensing (InSAR).

BOREHOLE EXTENSOMETERS—ONGOING MONITORING (2016–2022)

As shown in Figures 10.4–10.7, borehole extensometers are wells designed for measuring compaction or expansion of an aquifer system (Galloway et al., 1999). Extensometers typically are paired with monitoring wells so that correlation between groundwater-level changes and aquifer compaction can be determined.

In Hampton Roads from 1996 to 2016, no borehole extensometers were active. However, historic extensometer data are available, covering the period 1979 to 1995 for an extensometer located at Franklin, Virginia, and 1982 to 1995 for an extensometer located at Suffolk, Virginia (Pope and Burbey, 2004).

These older existing extensometers at Franklin and Suffolk have recently been equipped by the USGS with digital potentiometers, dial gauges, and satellite telemetry to provide aquifer compaction measurements with sub-millimeter (0.01 mm) accuracy. Data are being collected to test if the extensometer stations can be reactivated to detect aquifer compaction and expansion. The extensometers will be monitored for several months, and, if monitoring results are successful, the extensometers may be reactivated on a long-term basis. The possibility of installing GPS antennas on the extensometers, to determine contributions to subsidence from glacial isostatic rebound, will also be investigated (USGS, 2016).

Michelle Sneed, a USGS expert on subsidence and extensometers, was brought in to consult on land subsidence monitoring options in Hampton Roads. She described how, in California, extensometers provide the basis for understanding how land subsidence is related to groundwater withdrawals, for calibrating InSAR estimates of land subsidence, and for calibrating predictive models of land subsidence. Extensometers there provide data used for water-resource planning and subsidence-mitigation planning (USGS, 2016).

GEODETIC SURVEYING—ONGOING MONITORING (2016–2022)

Geodetic surveying is the measurement of land surface position. Global positioning system technology is now widely used to perform geodetic surveying. Permanent GPS stations, such as the network of Continuously Operating Reference Stations operated by the National Geodetic Survey (NGS), provide continuous information about land surface motion at single locations. CORS stations typically achieve centimeter-scale accuracy for absolute vertical position measurement and millimeter-scale accuracy for differential vertical position measurement. Permanent geodetic stations, such as CORS, also provide valuable information for calibrating remote sensing measurements of subsidence (USGS, 2016).

Survey networks consisting of multiple high-integrity monuments (benchmarks) that are installed on land periodically occupied with GPS antennas to measure land surface position can also provide valuable regional estimates of land subsidence. Dr. Philippe Hensel (NGS) has offered his expertise to help design and implement such a survey network for Hampton Roads.

A separate type of geodetic surveying that would be valuable for understanding land subsidence in Hampton Roads is using GPS antennas on bedrock wells to measure bedrock surface motion. This can be done at any new extensometer that is constructed. Existing bedrock wells, such as those at Franklin and Suffolk, may also be available as platforms for this type of monitoring.

The NGS, the lead US federal agency for surveying and geodetic science, operates the CORS network of benchmark stations that continuously record land surface positions in fine detail in three dimensions. The CORS network includes five benchmark sites in Hampton Roads.

Various other organizations have established continuous monitoring GPS antennas at benchmark stations in Hampton Roads that are not part of the CORS network. For example, the NASA Langley Research Center in Hampton, Virginia, established four benchmark sites with GPS antennas in 2015. In some cases, data from these non-CORS stations are available and, if a site has been constructed and operated following NGS guidelines (NGS, 2013; Floyd, 1978), the resulting data can be high quality and useful for subsidence calculation.

The NGS is currently (2017–2022) analyzing historic surveys of first-order benchmark sites on the Atlantic coast, including in Hampton Roads, to determine rates of subsidence over the past century. This study will produce maps of subsidence rates over multiple time periods.

TIDAL STATIONS—ONGOING MONITORING (2016–2022)

For decades, the National Oceanic and Atmospheric Administration (NOAA) has operated tidal stations to provide continuous water-level data at four sites in Hampton Roads. Data are publicly available at no cost from NOAA's website.

REMOTE SENSING—ONGOING MONITORING (2016–2022)

Interferometric Synthetic Aperture Radar is a remote sensing technique used to measure land surface elevation changes over wide areas, for example, over the entire Hampton Roads area. InSAR can be used to determine and map critical areas of land subsidence, select locations for detailed geodetic surveying, and provide strategies for preventing and mitigating land subsidence (Bawden et al., 2003). Accuracy of InSAR subsidence estimates will be important in Hampton Roads, because subsidence rates in the area have been measured at 1.1 to 4.8 millimeters, as compared to typical error for InSAR of 5–10 mm. The high atmospheric humidity and dense vegetation found in Hampton Roads can reduce InSAR accuracy. Problems with error can be overcome by analyzing a large number of satellite scenes, applying persist scatter analysis techniques, using INSAR data collected over multiple years, and using L-band or X-band rather than C-band InSAR data (USGS, 2016).

Probably the most valuable aspect of InSAR remote sensing is its capacity to input valuable data for detailed mapping of regional subsidence over time. InSAR data have been collected for Hampton Roads by various satellites since 1992 and are currently collected by several international satellites. In 2017, a new United States satellite, NISAR, began collecting InSAR data over Hampton Roads (USGS, 2016).

The Bottom Line: The U.S. Geological Survey is cooperating with federal, state, and local government agencies to study and better understand the problem of land subsidence in the southern Chesapeake Bay region. In order to make informed decisions, local resource managers, planners, politicians, and regulators need in-depth knowledge of the particulars involved with relative sea-level rise and land subsidence in the Hampton Roads area. This knowledge is necessary in the planning of increased flood risks and preventing land subsidence.

The Real Bottom Line: The intended purpose of this book is to satisfy the elements described previously—to explain what relative sea-level and land subsidence are and to describe and provide nuts-and-bolts explanations (in the chapters to follow) for planners, managers, and others, a methodology and technology that can potentially prevent, provide rebound from, or mitigate the impact of land subsidence in the Hampton Roads region.

HAMPTON ROADS SANITATION DISTRICT

Oysters drove its genesis. No, not the genesis of Chesapeake Bay; its genesis was driven by a heavy, unstoppable, all-knowing hand. Hampton Roads Sanitation District (HRSD), arguably the premier wastewater treatment district on the globe, became a viable governor-appointed state commission–monitored entity because of a significant decline in the oyster population in Chesapeake Bay. As a case in point, consider that in the Hampton Roads region of Chesapeake Bay in 1607, when Captain John Smith and his team settled in Jamestown, oysters up to 13 inches in size were plentiful—more than could ever be harvested and consumed by the handful of early settlers. And this population of oysters and other aquatic lifeforms remained plentiful until the population gradually increased in the Bay region.

Over-harvesting of oysters by the increased numbers of humans living in the Chesapeake Bay region was (and might still be) a major issue with the decline of the oyster population. However, the real culprit causing the decline in the oyster population is pollution. Before the Bay became polluted from sewage, sediment, and garbage disposal, oysters could manage natural pollution from stormwater run-off and other sources. Ninety years ago, when there was a much larger oyster population than today, it is estimated that the large oyster population could filter pollutants from the Bay and clean it in as little as four days. By the 1930s, however, the declining oyster population was overwhelmed by the increasing pollution levels.

For years, and in written accounts, the author has stated that pollution is a judgment call. That is, pollution as viewed by one person may not be pollution observed by another. You might shake your head and ask a couple of questions, "Pollution is a judgment call? Why is pollution a judgment call?" A judgment is based on an opinion; it is an opinion because people differ in what they consider a pollutant based on their assessment of the accompanying benefits and/or risks to their health and economic well-being posed by the pollutant. For example, visible and invisible chemicals spewed into the air or water by an industrial facility might be harmful to people and other forms of life living nearby. However, if the facility is required to install expensive pollution controls, forcing the industrial facility to shut down or move away, workers who would lose their jobs and merchants who would lose their livelihoods might feel that the risks from polluted air and water are minor weighed against the benefits of profitable employment and business opportunity. The same level of pollution can also affect two people quite differently. Some forms of air pollution, for example, might cause a slight irritation to a healthy person but cause life-threatening problems to someone with chronic obstructive pulmonary disease (COPD) like emphysema. Differing priorities lead to differing perceptions of pollution (concern at the level of pesticides in foodstuffs generating the need for wholesale banning of insecticides is unlikely to help the starving). No one wants to hear that cleaning up the environment is going to have a negative impact on them. The fact is public perception lags behind reality because the reality is sometimes unbearable.

First, along with the over-harvesting of oysters (including crabs and other species of Bay life), pollution has degraded the waters of Chesapeake Bay and directly contributed to the decrease in the oyster population therein. Second, pollution did not become an issue, like it did with the compost facility and paper mill, until the "neighbors" complained, and they did complain. Not only in the 1930s were they complaining about the reduced population of sea life in southern Chesapeake Bay (and other regions of the Bay), but the accumulation of floating sewage and the amplification of nasty odors emanating from the biodegradation of the sewage got passersby, would-be swimmers, and boaters and fisher-people's attention. Third, again, pollution is a judgment call, and in 1940 voters made the judgment to authorize the governor to appoint a representative commission to oversee pollution mitigation of the southern Chesapeake Bay region. Thus, Hampton Roads Sanitation District came into being. HRSD is a political subdivision of the Commonwealth of Virginia with a service area that includes 17 counties and cities encompassing 2,800 square miles in its southeastern Virginia service area (see Figure 10.8). HRSD's collection system consists of more than 500 miles of piping, 6–66 inches in diameter. HRDS possesses more than 100 active pumping operations that pump raw wastewater to nine major treatment plants in Hampton Roads and four smaller plants in the Middle Peninsula. The combined capacity of HRSD facilities is 249 million gallons per day (MGD).

Probably another question rumbling through the reader's brain matter at this point is "Has HRSD solved the pollution problem in Chesapeake Bay?" This question leads us to another question, "Have the oysters rebounded in quantity?" The answer to both these questions is yes—that is, to a point. On an ongoing, 24–7 basis, HRSD treats wastewater to a quality better than what is contained in the James River, Elizabeth River, York River, and other river systems in the region. Those who have no knowledge of wastewater treatment, HRSD, and/or the conditions of the rivers in this region might have second thoughts about this statement. However, it is true. It is all about the human-made water cycle. In this case, we are talking about the urban water cycle.

Water and wastewater professionals maintain a continuous urban water cycle (the hidden function) on a daily basis.

HRSD Service Area

A Political Subdivision of the Commonwealth of Virginia

Facilities include the following:

1. Atlantic, Virginia Beach
2. Chesapeake-Elizabeth, Va. Beach
3. Army Base, Norfolk
4. Virginia Initiative, Norfolk
5. Nansemond, Suffolk
6. Lawnes Point, Smithfield
7. County of Surry
8. Town of Surry

09. Boat Harbor, Newport News
10. James River, Newport News
11. Williamsburg, James City County
12. York River, York County
13. West Point, King William County
14. King William, KingWilliam County
15. Central Middlesex, Middlesex County
16. Urbanna, Middlesex County

Serving the Cities of
Chesapeake, Hampton,
Newport News, Norfolk,
Poquoson, Portsmouth, Suffolk,
Virginia Beach, Williamsburg and the
Counties of Gloucester,
Isle of Wight, James City,
King and Queen, King William,
Mathews, Middlesex, Surry* and York
*Excluding the Town of Claremont

FIGURE 10.8   HRSD service area.

DID YOU KNOW?

Artificially generated water cycles or the urban water cycles consist of (1) source (surface or groundwater), (2) water treatment and distribution, (3) use and reuse, and (4) wastewater treatment and disposition, as well as the connection of the cycle to the surrounding hydrological basins.

So let's get back to the part of the statement that ended with "—to a point." There is no doubt that Chesapeake Bay is cleaner and that the sea life, including oysters, is happier today because of the efforts of HRSD. The problem of making the Bay cleaner is compounded by two factors. First, there are more than 300 wastewater treatment plants that outfall treated water to the Chesapeake via its nine major river systems and numerous tributaries. These treatment plants, separate and isolated from HRSD's thirteen plants, do the best they can to treat wastewater to a cleaner product (effluent) than the influent they received from various sources. However, some of these 300+ other plants treat only to primary treatment levels, and thus their effluent is not as clean as secondary and tertiary plant effluent. Second, HRSD treats wastewater to a top-notch water quality level. However, treating wastewater to remove nutrients is a complicated and expensive undertaking. Biological and other nutrient removal technologies are available and in use in many locales, but the technology is expensive—expensive to the point where the treatment technology needed and used might overtax the ratepayers.

Earlier it was pointed out that the Chesapeake Bay occasionally suffers from dead zones due to algal blooms. *Algae bloom* is a phenomenon whereby excessive nutrients within the Bay cause an explosion of plant life that results in the depletion of the oxygen in the water needed by fish and other aquatic life. Algae bloom is usually the result of urban runoff (of lawn fertilizers, etc.). The potential tragedy is that of a "fish kill," where the Bay life dies in one mass execution.

Algal bloom and dead zones and the resulting fish kill events are a major issue, of course. However, when you add this problem to relative sea-level rise and land subsidence, it can be readily seen that the issues and problems with maintaining the health of the Chesapeake Bay and its inhabitants are multifaceted.

SWIFT: THE PROCESS

With regard to the problems with the Chesapeake Bay, land subsidence, and relative sea-level rise in the Hampton Roads region, HRSD has developed the innovative Sustainable Water Initiative For Tomorrow program (a work in progress; a decadal project). Do not confuse the acronym "swift" with the adjectives fast, speedy, rapid, hurried, immediate, or quick. SWIFT is a long-term project that is being developed on a timeline that is set for installation of the technical equipment and operational procedures with a completion date of 2030.

What is HRSD's SWIFT? SWIFT is a program to inject treated wastewater into the subsurface; specifically, it is designed to inject treated wastewater to drinking water quality into the Potomac Aquifer. Injection of water into the subsurface is expected to raise groundwater pressures, thereby potentially expanding the aquifer system, raising the land surface, and counteracting land subsidence occurring in the Virginia Coastal Plain. In 2016 a pilot project site was under construction at the HRSD Nansemond Wastewater Treatment Plant in Suffolk, Virginia, to test injection into the aquifer system. HRSD has asked USGS to prepare a proposal for installation of an extensometer monitoring station at the test site to monitor groundwater levels and aquifer compaction and expansion.

Problem

SWIFT is designed to counter land subsidence at various locations in the Hampton Roads area of southern Chesapeake Bay where land subsidence rates of 1.1 to 4.8 millimeters per year have been observed (Eggleston and Pope, 2013; Holdahl and Morrison, 1974).

Injection of treated wastewater (treated to drinking water quality) is expected to counteract land subsidence or raise land surface elevations in the region. Careful monitoring of aquifer system compaction and groundwater levels can be used to optimize the injection process and to improve fundamental understanding of the relation between groundwater pressures and aquifer-system compaction and expansion.

There is more to SWIFT than just arresting or mitigating land subsidence and relative sea-level rise in the Hampton Roads region. One of the additional goals of the project is to stop discharge of treated wastewater from seven of its plants. This would mean 18 million pounds a year less of nitrogen, phosphorus, and sediment out-falling into the bay. Assuming SWIFT works as designed, this is a huge benefit to the Chesapeake Bay in that it may help to prevent or reduce the formation of algal bloom dead zones. Not only would success as a result of treated wastewater injection benefit the Bay, but it would also be a huge benefit for the ratepayers at HRSD. To meet regulatory guidelines to remove nutrients from discharged treated wastewater would cost hundreds and millions of dollars and almost non-stop retrofitting at the treatment plants to keep up with advances in treatment technology and regulatory requirements. Another goal of HRSD's SWIFT project is to restore or restock potable groundwater supplies in the local aquifers. The drawdown of water from the groundwater supply has not only contributed to land subsidence but also to a reduction of water available for potable use.

HRSD's planned restocking of Hampton Roads' groundwater supply with injected wastewater treated to potable water quality is not without its critics. The critics state that HRSD's wastewater injection project would contaminate potable water aquifers. For the critics and others, this is where the so-called "yuck factor" comes into play. The yuck factor, in this particular instance, has to do with the thought that groundwater for consumptive use will be contaminated basically with toilet water. This is the common view of many of the critics who feel HRSD's SWIFT project is nothing more than direct reuse of wastewater; that is, a pipe-to-pipe connection of toilet water to their home water taps.

What the critics and others do not realize is that we are already using and drinking treated and recycled toilet water. As far as HRSD's SWIFT project contaminating existing aquifers with toilet water, it is important to point out that this water is to be treated (and already is at Nansemond Treatment Plant in Suffolk, Virginia) to drinking water quality—to drinking water quality is the key phrase here. This sophisticated and extensive train of unit drinking water quality treatment process—treated wastewater that HRSD general manager and several others drank right out of the process recently, and, by the way, they are doing just fine today, thank you very much—is discussed in detail later in the text. The bottom line: Statements about the yuck factor involved in drinking treated toilet water are grossly over-stated, as pointed out in Sidebar 10.1.

SIDEBAR 10.1   WASTEWATER YUCK FACTOR OVER-STATED

That great mythical hero, Hercules, arguably the world's first environmental engineer, was ordered to perform his fifth labor by Eurystheus to clean up King Augeas' stables. Hercules, faced with a mountain of horse and cattle waste piled high in the stable area, had to devise some method to dispose of the waste; he did. He diverted a couple of river streams to the inside of the stable area so that all the animal waste could simply be deposited into the river streams: Out of sight, out of mind. The waste simply flowed downstream. Hercules understood the principal point in pollution control technology that is pertinent to this very day and to this discussion; that is, *dilution is the solution to pollution.*

When people say they would never drink toilet water, they have no idea what they are saying. As pointed out in my textbook, *The Science of Water,* 3rd edition, the fact is we drink recycled wastewater every day. In Hampton Roads, for example, HRSD's wastewater treatment plants outfall (discharge) treated water to the major rivers in the region. Many of the region's rivers are sources of local drinking water supplies. Even local groundwater supplies are routinely infiltrated with surface water inputs, which, again, are commonly supplied by treated wastewater (and sometimes infiltrated by raw sewage that is accidentally spilled).

My compliments to Mr. Henifin, general manager of HRSD, who stated in a recent local newspaper article that he would be first to drink the treated wastewater effluent from the unit treatment processes at York River Treatment Plant (and, as mentioned, he did). My only contention with his statement is that because of Mother Nature's water cycle, the one we all learned about in grade school, we have been drinking toilet water all along. I have yet to find anything yucky about it or its taste.

****

The SWIFT project includes construction of an extensometer monitoring station with the ability to accurately measure land-surface elevations, bedrock-surface elevations, and changes in aquifer-system thickness. Monitoring of groundwater levels and aquifer-system elastic response will benefit operation of the wastewater injection system at the Nansemond Treatment Plant in Suffolk, Virginia, and provide guidance of future wastewater injection facilities for the SWIFT project.

Note to Reader

At the beginning of this book, it was pointed out that in order to understand the basics of HRSD's SWIFT initiative, it would be necessary to connect the dots. In this section of this chapter, the final dot is put in place—HRSD's process of converting sewage (wastewater) into drinking water, then injecting it back into the Potomac Aquifer. Right now, it is happening at the Nansemond Treatment Plant, which

HRSD is using as a research center to gather data about effects on water quality and geology. The goal is to inject around 1 million gallons a day, but the process has had a few stops and starts as adjustments are made. The information presented in the following is operational data as we know it today as we continue to learn and adjust.

SWIFT: THE PROCESS

Figure 10.9 presents a process flow block diagram of seven-unit processes (seven-step process) connected in a treatment train for the SWIFT Demonstration Facility at HRSD's Nansemond Treatment Plant in Suffolk, Virginia. In addition, a very brief fundamental explanation of each of the seven-unit processes is provided.

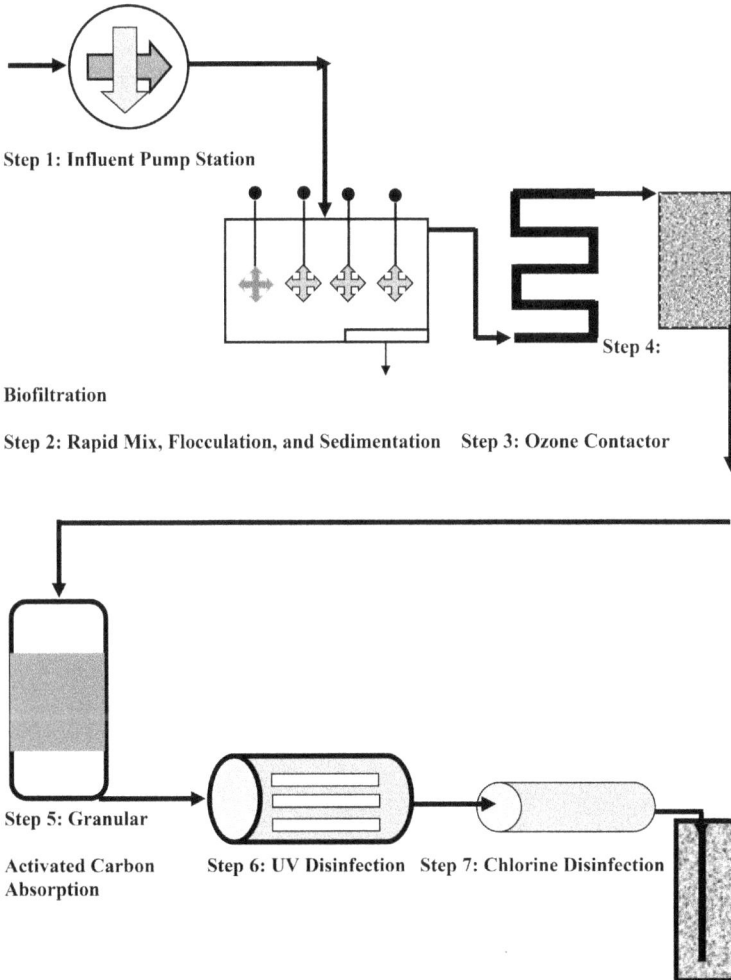

Step 1: Influent Pump Station

Step 4:

Biofiltration

Step 2: Rapid Mix, Flocculation, and Sedimentation   Step 3: Ozone Contactor

Step 5: Granular

Activated Carbon   Step 6: UV Disinfection Step 7: Chlorine Disinfection
Absorption

FIGURE 10.9   Process flow diagram for SWIFT demonstration facility.

SWIFT UNIT PROCESS DESCRIPTION

In this section, each of the SWIFT unit processes is described.

  Step 1: Influent pump station (see Figure 10.10). Wastewater effluent from the Nansemond Treatment Plant is directed to the SWIFT processing site on the plant location. Thus, the highly treated water from the Nansemond Treatment plant is pumped to the Research Center's advanced treatment facility (SWIFT), where it undergoes advanced treatment within the unit processes housed within; the highly treated SWIFT effluent (of better drinking water quality than that contained within the Potomac Aquifer) is outfalled (injected) into the Upper, Middle, and Lower Potomac Aquifer layers.

  Step 2: Mixing, flocculation, and sedimentation—This unit process removes suspended solids by settling large particles to the bottom of the water column.

  Step 3: Ozone Contact—This unit process breaks down organic material and provides disinfection.

  Step 4: Biologically Active Filtration—This unit process filters out suspended particles and pathogens and removes dissolved organic compounds through microbiological activity.

  Step 5: Granulated Activated Carbon Contactors—This unit removes trace organic compounds and prepares the water tor ultraviolet disinfection.

  Step 6: Ultraviolet (UV) Disinfection—This unit process provides a barrier to pathogens by disinfecting the water with high intensity ultraviolet light.

  Step 7: Chlorine Contact and Chemical Addition—This unit process provides disinfection of finished water using chlorine and serves as an additional barrier to pathogens. Chemical addition is used on the disinfected water and is adjusted by small doses to match the geochemistry of the water already more closely in the aquifer.

FIGURE 10.10   SWIFT influent pump station.

Source: Photo by F. R. Spellman.

THE ULTIMATE BOTTOM LINE

HRSD's SWIFT initiative is a far-thinking, innovative initiative that potentially offers enormous benefits not only for the Hampton Roads region but also for any region with similar needs, issues, and/or problems. To conclude with an ultimate bottom line on the SWIFT process and its benefits presented in this account, we must recognize the water challenges faced by those in the Hampton Roads region. These water challenges consist of questions that only operational time and adjustments can and will eventually answer. These questions are:

- Will SWIFT restore Chesapeake Bay?
- Will SWIFT mitigate groundwater depletion?
- Will SWIFT prevent saltwater intrusion?
- Will SWIFT counter relative sea level rise?
- Will SWIFT prevent recurrent flooding?
- Will SWIFT prevent sanitary sewer overflows?
- Will SWIFT be affordable?

The jury is still out on whether these questions will be answered in the positive. We do not know what we do not know at this precise moment. However, early, very early observations, measurements, and results indicate promise.

By the way, if you ever visit the SWIFT research center at the Nansemond Treatment Plant in Suffolk, Virginia, make sure to ask to sample the final product at the sample tasting location shown in Figure 10.11. Those who have sampled the

FIGURE 10.11   Hand pump used by site visitors to sample SWIFT water.

Source: Photo by F. R. Spellman.

recycled water have had nothing but positive comments and have noted how surprised they were that the SWIFT water tasted better than any water they have ever
consumed.

And the positive comments just related are the *real ultimate bottom line.*

NOTES

1. Definitions are based on USGS (1972) *Glossary of Selected Terms Useful in Studies of
   the Mechanics of Aquifer Systems and Land Subsidence due to Fluid Withdrawal* by J.F.
   Poland, B.E. Lofgren, and F.S. Riley. Washington, DC: United States Printing Office; and
   F. Spellman (2020) *The Science of Water*, 4th ed. Boca Raton, FL: CRC Press.
2. Adaptation from F. Spellman (2021) *Sustainable Water Initiative for Tomorrow (SWIFT)*.
   Lanham, MD: Bernan Press.
3. Material in this chapter is based on USGS (2013). *Land Subsidence in the United States*.
   Circular 1182. U.S. Department of the Interior, U.S. Geological Survey. Washington, DC.

REFERENCES

American Society of Civil Engineers. (1962). Nomenclature for hydraulics: Am. Soc. Civil
   Engineers, Manual and Repts, on Engineering Practice No. 43, p. 85.
Bawden, G.W., Sneed, M., Stork, S.V., and Galloway, D.L. (2003). Measuring human-induced
   land subsidence from space: U.S Geological Survey Fact Sheet 069–03, 4 p., http://pubs.
   usgs.gov/fs/fs/fs069003/.
Cahoon, D.R., Reed, D.J., Kolker, A.S., Brinson, M.M., Stevenson, J.C., Riggs, S., Christian,
   R., Reyes, E., Voss, C., and Kunz, D. (2009). Coastal wetland sustainability, chap. 4
   of Titus, J.G., Anderson, K.E., Cahoon, D.R., Gesch, D.B., Gill, S.K., Gutierrez, B.T.,
   Thieler, E.R., and Williams, S.J., editors. *Coastal Sensitivity to Sea-Level Rise—A
   Focus on the Mid-Atlantic Region, A Report by the U.S. Climate Change Science
   Program and the Subcommittee on Global Change Research*. Washington, DC: U.S.
   Environmental Protection Agency U.S. Climate Change Science Program Synthesis
   and Assessment Product 4.1, pp. 57–72.
Dawidowski, J.B., and Koolen, J.J. (1994). Computerized determination of the preconsoli-
   dation stress in compaction texting of field core samples. *Soil and Tillage Research*
   31(2):277–282.
Eggleston, J., and Pope, J. (2013). *Land Subsidence and Relative Sea-Level Rise in the
   Southern Chesapeake Bay Region*. USGS Circular 2013–1392. Reston, VA. Accessed
   12/21/16 @ http://dx.doi.org/10.3133/cir1392.
Erwin, R.M., Brinker, D.F., Watts, B.D., Costanzo, G.R., and Morton, D.D. (2011). Islands at
   bay—Rising seas, eroding islands, and waterbird habitat loss in Chesapeake Bay, USA.
   Journal of Coastal Conservation 15:51–60.
Federal Emergency Management Agency. (2002). Flood insurance study of Franklin, Virginia,
   community 510060 (revised September 4, 2002): Federal Emergency Management
   Agency, 16 p.
Floyd, R.P. (1978). *Geodetic Benchmarks, NOAA Manual NOS NGS 1*. Rockville, MD: U.S.
   Dept of Commerce, National Oceanic and Atmospheric Administration, September, 52
   pp. Accessed @ www.ngs.noaa.gov/PUBS_LIB/GeodeticBMs.pdf.
Galloway, D.L., and Hoffmann, J. (2007). The application of satellite differential SAR
   interferometry-derived ground displacements in hydrogeology. *Hydrogeology Journal*
   15(1):133–154.

Galloway, D.L., Jones, D.R., and Ingebritsen, S.E. (eds.) (1999). Land Subsidence in the United States: U.S. Geological Survey Circular 1182, 177 p. Accessed @ http://pubs.usgs.gov/circ/circ1182/.

Helm, D.C. (1975). One-dimensional simulation of aquifer system compaction near Pixley, Calif., part 1. *Constant Parameters: Water Resource Research* 11:465–478.

Holdahl, S.R., and Morrison, N., (1974). Regional investigations of vertical crustal movements in the U.S., using precise relevelings and mareograph data. *Tectonophysics* 23(4):373–390.

Kirwan, M.L., and Guntenspergen, G.R. (2012). Feedbacks between intimidation, root production, and shoot growth in a rapidly submerging brackish marsh. *Journal of Ecology* 100(3):760–770.

Kirwan, M.L., Langley, J.A., Guntenspergen, G.R., and Megonigal, J.P. (2012). The impact of sea-level rise on organic matter decay rates in Chesapeake Bay brackish tidal marshes. *Biogeosciences Discussions* 9(10):14689–14708.

McFarlane, B.J. (2012). Climate change in Hampton Roads—Phase III—Sea level rise in Hampton Roads, Virginia: Chesapeake, Virginia, Hampton Roads Planning District Commission report PET13–06, July, 102 p. du/Grey Lit/VIMS/sramsoe425.pdf.lus maps. Accessed @ www.hrpdeva.gov/uploads/docs/HRPDC_ClimateChangeReport2012_Rull_Reduced.pdf.

Morris, J.T., Sundareshwar, P.V., Nietch, C.T., Kjerfve, B., and Cahoon, D.R. (2002). Responses of coastal wetlands to rising sea level. *Ecology* 83(10):2869–2877.

National Geodetic Survey. (2013). Guidelines for new and existing Continuously Operating Reference Stations (CORS) National Geodetic Survey National Ocean Survey, NOAS Silver Spring, Maryland, January. Accessed @ http://ngs.noaa.gov?PUBS_LIB/CORS_grfuidlines.pdf.

Poland, J.F. (ed.) (1984). *Guidebook to Studies of Land Subsidence Due to Ground-Water Withdrawal*. United Nations Educational, Scientific and Cultural Organization, 305 p. plug appendices. Accessed @ wwwrcamni.wr.usgs.go/rgws/Unesco/PDF-Cahpters/Guidebook.pdf.

Poland, J.F., and Green, J.H. (1962). Subsidence in the Santa Clara Valley, California: A progress report. U.S. Geological Survey Water-Supply Paper 1619-C, 16 p.

Poland, J.F., and Ireland, R.L. (1988). Land subsidence in the Santa Clara Valley, California, as of 1982. U.S. Geological Survey Professional Paper 497-F, 61 p.

Pope, J.P., and Burbey, T.J. (2004). Multiple-aquifer characteristics from single borehole extensometer records. *Ground Water* 42(1):45–58.

Pritchard, M.E. (2006). InSAR, a tool for measuring Earth's surface deformation. *Physics Today* 59(7), July:68–69.

Sallenger, A.H., Doran, K.S., and Howd, P.A. (2012). Hotspot of accelerated sea-level rise on the Atlantic coast of North America. *Nature Climate Change* 2(12):884–888.

Snay, R.A., and Soler, T. (2008). Continuously Operating Reference Station (CORS): History, applications, and future enhancements. *Journal of Surveying Engineering* 134(4), November 1:95–104.

Sneed, M., Stork, S.V., and Ikehara, M.E. (2002). Detection and measurement of land subsidence using global position system and interferometric synthetic aperture radar, Coachella Valley, California, 1998–2000: U.S. *Geological Survey Water-Resources Investigations Report*:02–4239, 29 p.

Stork, S.V., and Sneed, M. (2002). Houston-Galveston Bay area, Texas, from space: A new tool for mapping land subsidence: U.S. Geological Survey Fact Sheet 2002–110, 6 p. Accessed @ http://pubs.usgs.gov/fs/fs-110-02/.

Sweet, W., Park, J., Marra, J., Zervas, C., and Gill, S. (2014). Sea level rise and nuisance flood frequency changes around the United States, NOAS Tech. Report NOS CO-OPS 073. 58 pp. Accessed @ http://tidesandcurrents.noaa.gov/publications/NOAA_Technical_Report_NOS_COOP_073.pdf.

Terzaghi, K. (1925). Principles of soil mechanics, IV—Settlement and consolidation of clay. *Engineering New-Record* 95(3):874–878.

Terzaghi, K., and Peck, R.B. (1948). *Soil Mechanics in Engineering Practice.* New York: John Wiley and Sons, Inc., 566 p.

Tolman, C.F., and Poland, J.F. (1940). Ground-water infiltration, and ground-surface recession in Santa Clara Valley, Santa Clara County, California. *Transactions American Geophysical Union* 21:23–24.

USGS. (2013). *Land Subsidence in the United States.* Circular 1182. U.S. Department of the Interior, U.S. Geological Survey. Washington, DC.

USGS. (2016). Land subsidence monitoring in Hampton Roads: Progress report. Jack Eggleston, Virginia Water Science Center. Accessed @ www.hrpdc.gov/uploads/docs/04A_attachment_USGS_HRPDC_Subsi . . .

USGS. (1997). Location of Chesapeake Bay. Accessed @ http://woodshole.er.usgs.gov/epubs/bolide/location_of_bay.html.

USGS. (1998). *The Chesapeake Bay Bolide.* Washington, DC: US Geological Survey.

Zervas, C. (2009). Sea level variations of the United States, 1854–2006: National Oceanic and Atmospheric Administration Technical Report NOS CO-OPS 053, 76 p. plus appendixes. Accessed @ www.co-ops.nos.noas.gov/publicatins/Tech_r[r_53.pdf.

SUGGESTED READINGS

Editorial: A Worrisome, Watery Future. (2022, February 19). *The Virginian Pilot.* Accessed @ www.pilotonline.com/opinion/editorials/vp-ed-edtorial-climate-change-hampton-roads-0220-20220219-hpjrwmexyzb7njlj7o2nyatgae-story.html.

Poland, J.R., Lofgren, B.E., Ireland, R.L., and Pugh, R.G. (1975). Land subsidence in the San Joaquin Valley, California, as of 1972. U.S. Geological Survey Professional Paper 437-H, 78 p.

Rancoules, D., Bourgine, B., de Michele, M., Le Cozannet, G., Closset, L., Bremmer, C., Veldkamp, H., Tragheim, D., Bateson, I., Crosetto, M., Agudo, M., and Engdahl, M. (2009). Validation and intercomparison of persistent scatterers interferometry—PSIC4 project results. *Journal of Applied Geophysics* 68(3):335–347.

USGS. (2016). *The Chesapeake Bay Bolide Impact: A New View of Coastal Plain Evolution.* Fact Sheet 049–98. Accessed @ http://pubs.usgs.gov/fs/fs/49-98/.

11 When Land Fractures

It's been raining a lot, or very hot—it must be earthquake weather!

INTRODUCTION

Anyone who has witnessed (been exposed to—during shake, rattle, roll, and/or total destruction), studied, or looked upon a scene like the ones shown in Figures 11.1 and 11.2 recognizes that earthquakes are unforgettable events—possibly life-changing or life-ending events. Again, truth be told, with over a million or so earthquakes occurring each year on Earth, it is unlikely those exposed to such events will ever forget such occurrences. Even though most earthquakes are insignificant, a few thousand of these produce noticeable effects such as tremors and/or ground shaking. The passage of time has shown that about 20 earthquakes each year cause major damage and destruction—ground shaking, surface faulting, ground failure, and less commonly tsunamis. It is estimated that about 10,000 people die each year because of earthquakes.

Okay, it is likely that those who have read the beginning of this introduction to earthquakes will not or would not have any problem with agreeing with what has just

FIGURE 11.1   Searles Valley 6.4 earthquake that struck the Ridgecrest, California, area July 4, 2019.

Source: Public Domain. Photo by USGS/Ben Brooks.

DOI: 10.1201/9781003295211-11   **149**

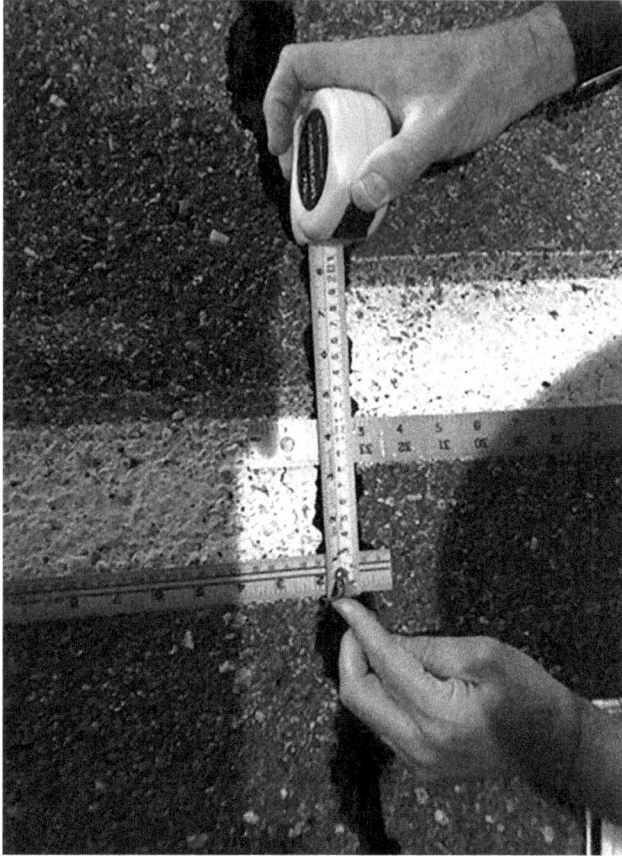

FIGURE 11.2   Measuring earthquake damage (6.4 quake) in Searles Valley, California, on Highway 178, July 4, 2019.

Source: USGS. Public domain photo.

been stated. But this is a book about elements of climate change being a driver of migration or displacement. So are earthquakes drivers of migration and displacement of living organisms on Earth?

This is the question. In a feature article by NASA on Global Climate Change, October, 20, 2019, *Can Climate Affect Earthquakes, or Are the Connections Shaky?* NASA details the various climate change occurrences (drivers), including warming temperatures, increased storm intensity, droughts, fire and ice, human-caused changes, and so forth, but does not definitively note climate change as a trigger of earthquakes. However, the jury is still out on the possibility of climate change elements causing earthquakes. On the other hand, there is evidence (and a lot of speculation) that climate change does interact with earth systems to trigger earthquakes. One of the best arguments for the causal relationship between climate change and earthquakes is put forward by Chi-Ching Liu of the Institute of Earth Sciences at

Taipei's Academia Sinica. In a paper published in the journal *Nature*, Liu and his colleagues (2009) provided evidence for typhoons causing slow-moving earthquakes. Liu and his colleagues point out that "an earthquake fault that is ready to go is like a coiled-spring—all that is need is the pressure of a handshake."

In the distant past (4th century B.C.), Aristotle proposed that earthquakes were caused by winds trapped in subterranean caves. The thinking was that small tremors were probably caused by air pushing on the cavern roofs and larger ones by the air actually breaking the surface. This belief led to a belief in earthquake weather: that because a large amount of the air was trapped underground, the weather would be hot and calm before an earthquake. Note that a later theory stated that earthquakes occurred in calm, cloudy conditions and were usually preceded by strong winds, fireballs, and meteors.

According to USGS (2022), there is no such thing as "earthquake weather." Statistically, there is approximately an equal distribution of earthquakes in cold weather, hot weather, rainy weather, and so forth. Very large low-pressure changes associated with major storm systems (typhoons, hurricanes, and so forth) are known to trigger episodes of fault slip (slow earthquakes) in the Earth's crust and may also play a role in triggering damaging earthquakes—this is the point Liu and his colleagues make in their journal article. And it is this possibility that necessitates the inclusion of earthquakes as being generated, even in small insignificant numbers, to be included in this narrative.

The bottom line: this book suggests that climate change "could be" a causal factor involved with earthquakes. Keep in mind that the key words here are "could be"—and because it is possible but yet to be proved definitively, it can be said that someday, like the absolute cause of cancer being determined, the truth will be known about climate change and earthquake generation.

For the sake of this book, we state that climate change may cause earthquakes, and therefore the subject matter is worth discussing to the degree possible.

What Causes Earthquakes?

What is possible to state at the present time is that catastrophic earthquakes displace people and animals. We can also discuss causes of earthquakes—causes that we know for certain. Also, keep in mind that earthquakes can be devastating to the nth degree and not only drive local inhabitants—animals and human beings—to be displaced and to migrate to what are deemed safer locations (the so-called greener pastures) but also can kill and injure.

So let's talk about earthquakes.

Over the millennia, the effect of damaging earthquakes has been obvious to those who witnessed the results. However, the *cause* of earthquakes has not been as obvious. For example, the cause of earthquakes has initially shifted from the blaming of super-incantations of mythical beasts, to the wrath of Gods, to unexplainable magical occurrences, and/or to just normal natural phenomena occasionally required to retain Earth's structural integrity; that is, providing Earth with a periodic form of feedback to keep "things" in balance. We can say, overall, that an earthquake on Earth provides our planet with a sort of a geological homeostasis needed to maintain life as we know it.

Through the ages earthquakes have also come under the attention and eventually the pen of the world's greatest writers. Consider, for example, Voltaire's classic satirical novel, *Candide*, published in 1759, in which he mercilessly satirizes science and, in particular, earthquakes. Voltaire based the following comments on the 1755 Great Lisbon, Portugal, earthquake, to which the deaths of more than 60,000 people were attributed.

Dr. Pangloss says to Candide (on viewing the total devastation of Lisbon):

> the heirs of the dead will benefit financially; the building trade will enjoy a boom. Private misfortune must not be overrated. These poor people in their death agonies, and the worms about to devour them, are playing their proper and appointed part in God's master plan.

Although we still do not know what we do not know about earthquakes and their causes, we have evolved from using witchcraft or magic to explain their origins to the scientific methods employed today. In the first place, we do know that earthquakes are caused by the sudden release of energy along faults. Earthquakes are usually followed by a series of smaller earthquakes that we call aftershocks. Aftershocks represent further adjustments of rock along the fault. There are currently no reliable methods for predicting when earthquakes will occur.

With regard to the cause(s) or origin(s) of earthquakes, we have developed a couple of theories. One of these theories explains how earthquakes occur via *elastic rebound*. That is, according to elastic rebound theory, subsurface rock masses subjected to prolonged pressures from different directions will slowly bend and change shape. Continued pressure sets up strains so great that the rocks will eventually reach their elastic limit and rupture (break) and suddenly snap back into their original unstrained state. It is the snapping back (elastic rebound) that generates the seismic waves radiating outward from the break. The greater the stored energy (strain), the greater the release of energy.

The coincidence of many active volcanic belts with major belts of earthquake activity (*seismic* and *volcanic activity*) indicate that volcanoes and earthquakes may have a common cause. Plate interactions commonly cause both earthquakes (tectonic earthquakes) and volcanoes.

SEISMOLOGY

Even though *seismology* is the study of earthquakes, it is actually the study of how seismic waves behave in the Earth. The source of an earthquake is called the *hypocenter* or *focus* (i.e., the exact location within the Earth where seismic waves are generated). The *epicenter* is the point on the Earth's surface directly above the focus. Seismologists want to know where the focus and epicenter are located so a comparative study of the behavior of the earthquake event can be made with previous events—in an effort to further understanding.

Seismologists use instruments to detect, measure, and record seismic waves. Generally, the instrument used is the *seismograph*, which has been around for a long time. Modern updates have upgraded these instruments from the paper or magnetic tape strip to electronically recorded data that is input to a computer. A study of the relative arrival times of the diverse types of waves at a single location can be used to determine the distance to the epicenter. To determine the exact epicenter location, records from at least three widely separated seismograph stations are required.

SEISMIC WAVES

As mentioned, a portion of the energy released by an earthquake travels through the Earth. The speed of a seismic wave depends on the density and elasticity of the materials through which they travel. Seismic waves come in three main types:

- **P-Waves**—Primary, pressure, or push-pull waves (arrive first—first detected by seismograph) are compressional waves (expand and contract) that travel through the earth (solids, liquids, or gases) at speeds of from 3.4 to 8.6 miles per second. P waves move faster at depth, depending on the elastic properties of the rock through which they travel. P waves are the same thing as sound waves.
- **S-Waves**—Secondary or shear waves travel with a velocity (between 2.2 and 4.5 miles per second) that depends only on the rigidity and density of the material through which they travel. They are the second set of waves to arrive at the seismograph and will not travel through gases or liquids; thus, the velocity of S-waves through gas or liquids is zero.
- **Surface Waves**—Several types travel along the Earth's outer layer or surface or on layer boundaries in the earth. These are rolling, shaking waves that are the slowest waves but the ones that do the damage in large earthquakes.

EARTHQUAKE MAGNITUDE AND INTENSITY

The size of an earthquake is measured using two parameters—energy released (magnitude) and damage caused (intensity).

Earthquake Magnitude

The size of an earthquake is usually given in terms of its Richter magnitude. Richter magnitude is a scale devised by Charles Richter that measures the amplitude (height) of the largest recorded wave at a specific distance from the earthquake. A better measure is the Richter scale, which measures the total amount of energy released by an earthquake as recorded by seismographs. The amount of energy released is related to the Richter scale by the equation:

$$\text{Log } E = 11.8 + 1.5 \text{ M}$$

where
Log = the logarithm in base 10
E = the energy released in ergs
M = the Richter magnitude

In using the equation to calculate Richter magnitude, it quickly becomes apparent that each increase in 1 in Richter magnitude yields a 31-fold increase in the amount of energy released. Thus, a magnitude 6 earthquake releases 31 times more energy than a magnitude 5 earthquake. A magnitude 9 earthquake releases 31 × 31 or 961 times more energy than a magnitude 7 earthquake.

DID YOU KNOW?

While it is correct to say that for each increase in 1 in the Richter magnitude, there is a tenfold increase in amplitude of the wave, it is incorrect to say that each increase of 1 in Richter magnitude represents a tenfold increase in the size of the earthquake.

Earthquake Intensity

Earthquake intensity is a rough measure of an earthquake's destructive power (i.e., size and strength—how much the earth shook at a given place near the source of an earthquake). To measure earthquake intensity, Mercalli in 1902 devised an intensity scale of earthquakes based on the impressions of people involved, movement of furniture and other objects, and damage to buildings. The shock is most intense at the epicenter, which, as noted earlier, is located on the surface directly above the focus.

Mercalli's intensity scale uses a series of numbers (based on a scale of 1 to 12) to indicate different degrees of intensity (see Table 11.1). Keep in mind that this scale is somewhat subjective, but it provides a qualitative, but systematic, evaluation of earthquake damage.

TABLE 11.1
Modified Mercalli Intensity Scale

Intensity	Description
I	Not felt except under unusual conditions
II	Felt by only a few on upper floors
III	Felt by people lying down or seated
IV	Felt by indoors by many, by few outside
V	Felt by everyone, people awakened
VI	Trees sway, bells ring, some objects fall
VII	Causes alarm, walls, and plaster crack
VIII	Chimneys collapse, poorly constructed buildings seriously damaged
IX	Some houses collapse, pipes break
X	Ground cracks, most buildings collapse
XI	Few buildings survive, bridges collapse
XII	Total destruction

INTERNAL STRUCTURE OF EARTH

Information obtained from seismographs and other instruments indicates that the lithosphere may be divided into three zones: the crust, mantle, and core (see Figure 11.1).

Earth's Crust: The outermost and thinnest layer of the lithosphere is called the crust. There are two distinct types of crust: thin (as little as 4 miles in places) oceanic crust (composed primarily of basalt) that underlies the ocean basins and thicker continental crust (primarily granite 20 to 30 miles thick) that underlies the continents.

Earth's Mantle: Beneath the crust is an 1,800-mile-thick intermediate, dense, hot zone of semi-solid rock called the mantle. It is thought to be composed mainly of olivine-rich rock.

Earth's Core: Earth's core is about 4,300 miles in diameter. It is thought to be composed of a very hot, dense iron and nickel alloy. The core is divided into two different zones. The outer core is a liquid because the temperatures there are adequate to melt the iron-nickel alloy. The highly pressurized inner core is solid because the atoms are tightly crowded together.

PLATE TECTONICS

Within the past 45 or 50 years, geologists have developed the theory of plate tectonics (tectonics: Greek, "builder"). The theory of plate tectonics deals with the formation, destruction, and large-scale motions of great segments of Earth's surface (crust), called *plates*—this area of discussion is important with regard to earthquakes and

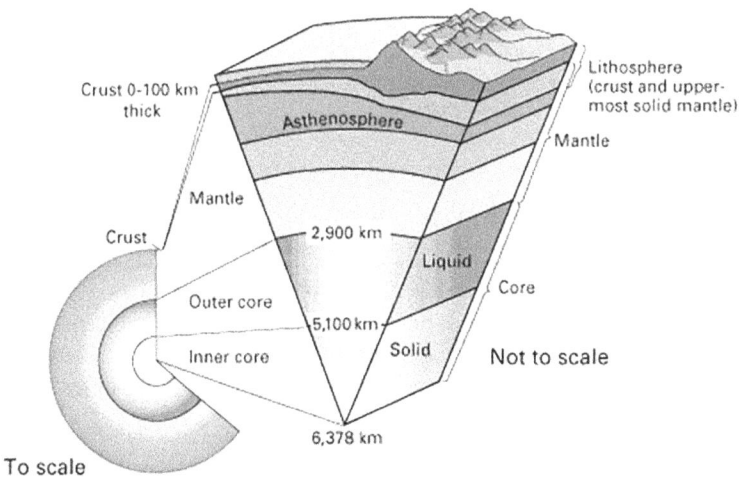

FIGURE 11.3 Structure of earth.

Source: From USGS (1999). Accessed 08/05/2022 @ https://pubs.usgs/gip/dynamic/inside.htm.

other Earth functions. This theory relies heavily on the older concepts of continental drift (developed during the first half of the 20th century) and seafloor spreading (understood during the 1960s), which help to explain the cause of earthquakes and volcanic eruptions and the origin of fold mountain systems.

CRUSTAL PLATES

Earth's crustal plates are composed of great slabs of rock (lithosphere), about 100 km thick, that cover thousands of square miles (they are thin in comparison to their length and width); they float on the ductile asthenosphere, carrying both continents and oceans. Generally, geologists recognize at least eight main plates and several smaller ones. These *main* plates include

- African Plate covering Africa—Continental plate
- Antarctic Plate covering Australia—Continental plate
- Australian Plate covering Australia—Continental plate
- Eurasian Plate covering Asia and Europe—Continental plate
- Indian Plate covering Indian subcontinent and a part of Indian Ocean—Continental plate
- Pacific Plate covering the Pacific Ocean—Oceanic plate
- North American Plate covering North America and northeast Siberia—Continental plate
- South American Plate covering South America—Continental plate

The *minor* plates include

- Arabian Plate
- Caribbean Plate
- Juan de Fuca Plate
- Cocos Plate
- Nazea Plate
- Philippine Plate
- Scotia Plate

Plate Boundaries

As mentioned, the asthenosphere is the ductile, soft, plastic-like zone in the upper mantle on which the crustal plates ride. Crustal plates move in relation to one another at one of three types of plate boundaries: convergent (collision boundaries), divergent (spreading boundaries), and transform boundaries. These boundaries between plates are typically associated with deep-sea trenches, large faults, fold mountain ranges, and mid-oceanic ridges.

Convergent Boundaries

Convergent boundaries (or active margins) develop where two plates slide towards each other, commonly forming either a subduction zone (if one plate subducts or moves underneath the other) or a continental collision (if the two plates contain continental crust). To relieve the stress created by the colliding plates, one plate is deformed and slips below the other.

Divergent Boundaries

Divergent boundaries occur where two plates slide apart from each other. Oceanic ridges, which are examples of these divergent boundaries, are where new oceanic, melted lithosphere materials well up, resulting in basaltic magmas which intrude and erupt at the oceanic ridge, in turn creating new oceanic lithosphere and crust (new ocean floor). Along with volcanic activity, the mid-oceanic ridges are also areas of seismic activity.

Transform Plate Boundaries

Transform, or shear/constructive, boundaries do not separate or collide; rather, they slide past each other in a horizontal manner with a shearing motion. Most transform boundaries occur where oceanic ridges are offset on the sea floor. The San Andreas Fault in California is an example of a transform fault.

REFERENCES

Chi-Ching Lui, Linde, A.T., and S, I.S. (2009). Slow earthquakes triggered by typhoons. *Nature* 460(7252).

SUGGESTED READINGS

Atkinson, L., and Sancetta, C. (1993). Hail and farewell. *Oceanography* 6(34).

Holmes, A. (1978). *Principles of Physical Geology* (3rd ed.). New York: John Wiley & Sons.

Lyman, J., and Fleming, R.H. (1940). Composition of seawater. *J Mar Res* 3:134–146.

McKnight, T. (2004). *Geographica: The Complete Illustrated Atlas of the World*. New York: Barnes and Noble Books.

Oreskes, N. (ed.) (2003). *Plate Tectonics: An Insider's History of the Modern Theory of the Earth*. New York: Westview.

Stanley, S.M. (1999). *Earth System History*. New York: W.H. Freeman, pp. 211–222.

Sverdrup, H.U., Johnson, M.W., and Fleming, R.H. (1942). *The Oceans: Their Physics, Chemistry, and General Biology*. New York: Prentice-Hall.

Turcotte, D.L., and Schubert, G. (2002). *Geodynamics: Second Edition*. New York: John Wiley & Sons.

USGS. (1981). What are the effects of earthquakes? Accessed 02/28/2022 @ www.usgs.gov/programs/eqrthquke-hazards/what=are=effects-earthquakes.

USGS. (2008). Earthquake hazards program. Accessed 09/14/2021 @ http://earthquakes.usgs.gov/.

USGS. (2022). Is there earthquake weather? Accessed 2/28/22 @ www.usgs.gov/fags/earthaqukes-weather.

Voltaire. (1991). *Candide*. New York, NY: Dover Publishing Inc. Mineola.

12 Climate Change and the Hispanic Paradox

TO THIS POINT

So far in this book, we have discussed and described climate change environmental drivers of migration for all types of lifeforms. These drivers include extreme weather events, sea level rise, coastal erosion, land subsidence, loss of agricultural productivity, water shortages (drought), increased pollution, habitat destruction, shifting vegetative patterns, and earthquake destruction. These certainly are drivers that make people pull up stakes, so to speak, and move on to those so-called greener pastures. Of course, there are other drivers of migration, but most of those have to do with societal issues like economics, poverty, slavey, brutality, personal attitudes, or political aspects—none of these drivers are addressed in this book, on purpose, because this is not, as stated before, a political book. Instead, this is a book about environmental changes that drive people from their native lands to other lands, hopefully to more environmentally hospitable and productive lands.

Keeping in line with those environmental drivers of migration, the one substantial area which has not been addressed to this point is disease and pestilence caused by climate change. Certainly, if an area, region, state, or country is changed via environmental changes, and these changes have a direct health aspect tagging along with the change, then it is not difficult to assume that people and other life forms will seek healthier environments to move to. Simply, if any of the climatic change elements present themselves in a particular area, then it may be impossible for the inhabitants, human and otherwise, to live there anymore—so they move on.

Before we move on to the health perspectives of climate change and its impact as a disease-causer that becomes a driver of displacement and migration, we set the stage with a discussion of our endangered atmosphere.

SETTING THE STAGE[1]

THE ENDANGERED ATMOSPHERE: CLIMATIC CHANGE

Humanity is conducting an unintended, uncontrolled, globally pervasive experiment whose ultimate consequences could be second only to nuclear war. The Earth's atmosphere is being changed at an unprecedented rate by pollutants resulting from human activities, inefficient and wasteful fossil fuel use, and the effects of rapid population growth in many regions. These changes are already having harmful consequences over many parts of the globe (Toronto Conference statement, June 1988).

Is climate change real? Is climate change occurring now? Are humans responsible for any climate change results, effects, or causes? Is climate change new?

DOI: 10.1201/9781003295211-12

Well, to answer the question "Is climate change new?", let's go back in time a bit and check out then compared to now.

Time: 10,312 BCE

He sat on the ground leaning against a deadfall, his leather-wrapped legs pulled tight under him, and watched the swamp. He felt disoriented, detached from the world around him. Even the air around him felt strange; it was different; this place was unusually warm. A possum waddled from a copse of vine-maple below him, and Yurk watched the possum move off to the left—the possum in a hurry, constantly jerking his head to the right, over his shoulder. The possum darted toward the marshy bank and stopped to sniff the ground. Some noise, or an odor carried on the wind, seemed suddenly to startle the possum into attention, and he looked back toward Yurk, then moved off into the tall marsh grass, where he disappeared from view.

The rain was coming down in a fine drizzle. The wind sighed through the fir boughs, and the afternoon was redolent with the smell of tree-perfumed air. Even with the light rain and wind, though, Yurk was warm—warmer than he ever remembered being before. He had never been so warm, his whole body at the same time. By a fire, only what faces the fire is warm.

Yurk's weathered face wore a mesmerized look as he chewed on a piece of bark, resting against the decaying trunk of the fallen tree, almost as if he were unaware of his surroundings. His eyes glazed over, as if he were there in shell form only—an empty one at that. Maybe his blank state and hypnotized appearance were the result of the view in front of those blank eyes: great truncated tree trunks blackened by fire stood above the surface of the swamp water, stark remnants of a very ancient past. A misty pall hung over the swamp as the blackness over the forest and the swamp took on an eerie, forbidding, spectral quality with the coming of night.

The cry of an owl drifted through the dark forest as Yurk stood (an effort that required much exertion from his tired, ancient body). Carefully he stretched and yawned—careful not because of his frailty, or from a sense of impending danger, but because of instinct—not fear exactly, just instinct. A lifetime, generations of lifetimes of vigilance for survival (both conscious and unconscious) had taught Yurk to be vigilant at all times in this place and time. He was leg-weary and footsore, but that really didn't concern him. He knew his ending time was near—that was why he had traveled better than 200 miles to this place. This place he had come to had been familiar to him years before, but in a very different form. He wanted to see the wonderment of the swampy terrain that lay before him now.

Yurk was viewing something he had heard about from other clan members but something he had never witnessed before. A swamp.

Yes, a swamp . . . with blackened, truncated tree remnants. In all his years (unusually old for his time and circumstances, easily more than 60), Yurk had never seen such a sight. Before—up until now—the landscape he had been familiar with had been covered in snow and ice. He had visited this place many times in the past—what seemed a bare plain of ice and snow. He had not been on this journey in many years, but the last time he had come, he had simply trudged through the open area (the swamp) over a bridge of thick ice and snow. He (and no one else) had any idea that the swamp lay below the thick layers of ice and snow. In his absence from this place, he had heard the tales from the younger clan hunters and had decided to take his last journey—to see such a place, such a site—before he died.

It was so warm.

As he stood, wiping his wet brow and looking out upon the swamp, thirty feet to Yurk's left, working toward the top of the steep, craggy ledge on the sheer cliff edge climbed the cat.

Like Yurk, the cat had come to this place many times in the past, although she could not cognitively determine the exact difference between the past and the present—she, too, knew this place had changed.

It was so warm.

In the past fifteen or so years, the cat, along with her running mates (these cats almost always ventured into the wilderness accompanied—to hunt and to kill required help—sometimes lots of help) had, like Yurk and his clan members, crossed the swamp using the ice-bridge. But now things were different; the cat knew this. She also knew that something else was different; it was so warm.

The cat (known today as smilodon or saber-toothed tiger) continued slowly, inexorably up the steep slope of the stony ridge. Unlike climbing to this ledge in the past, when she had allowed for the slipperiness of the ice sheet that covered the ledge, she should have had very little difficulty climbing the high terminal edge, overlooking the swamp. But now things were different—much different. She was on her last legs, in all ways. But her difficulty was even more than that; even though the going was easier now without the ice and snow, she still struggled her way up to the terminal point . . . it was so hot. She labored even to breathe.

Yurk and the cat were aware of each other. Each knew the other was there—have no doubt about that. Yurk probably more fearful of the cat than she was of him . . . but how could anyone tell? They had been bitter enemies throughout their lives. The cat preferred feasting on mammoths and mastodons (Yurk liked that kind of meat himself), but when confronted with her "only" threat, her only true enemy, the cat knew she was wise to be ready. Life of any sort was difficult enough . . . not being alert and wary at all times was certainly an invitation to disaster—for both of them.

It was so warm.

But now things were different. Neither the cat nor Yurk was attentive to each other; they were not as alert, as wary of each other as they had been in the past. Each knew, in their own way, that the days of hunting and protecting themselves were behind them—food certainly wasn't a consideration with either. No, food was not a problem; they were not hungry. Afraid? No, not really.

The cat continued her climb and finally reached the summit. She stood looking out upon the swamp (with one eye semi-focused on Yurk). Yurk stood below, looking out on the swamp (aware of her presence as well).

They both knew, in their own way, that things were different. Hell, they could feel the difference; it was so warm.

Warm . . . yes, it was warm. For their entire lives, they had never known such warmth, had never seen the snow and ice melt, had never witnessed the swampy landscape now before them. Their world was different . . . fearfully and wonderfully changed.

The warming trend had actually begun about two or three years earlier, though Yurk and the cat had barely been aware of it, because the increase in temperatures had been subtle . . . just about a half a degree Fahrenheit every three months or so. But now, now the difference was obvious. The temperature was a least ten degrees warmer than they had ever experienced—thus the melt, the freshly uncovered swamp, the rock-strewn ledge . . . and the warmth, of course.

The cat and Yurk stood for a time, gazing out at the swamp. What this change would mean to their clan and mates—those to follow—they were not capable of determining. What this change would bring to their world, they were not capable of speculating. So they stood, until Yurk sat back down on the ground, his back against the deadfall, and the cat lay down on the heated rocks of the ledge; they were both exhausted, tired, worn out—old, so old . . . and warm—too warm.

About an hour later, as darkness fell total upon the blackened, spectral landscape before them, they both went to sleep . . . the sleep of the dead . . . and their own warmth turned cold.

The ambient temperature continued to rise, even now that it was dark, night. A night that, when it ended, would bring the dawn of a new day . . . and the dawn of a new era.

It was so warm . . . and getting warmer.

The question is—Can't we do it with all the wastes we produce?

We think that goal is possible—especially if the scientific, political, social, and monetary commitment that needs to be made is made . . . and it will be.

Why?

What other choice do we have?

—(Spellman, 1999)

Note: The preceding account about Yurk, the saber-toothed cat (*Smilodon californicus*), and their surroundings can have different meanings, implications, or suggestions depending on point of view and the choice of meaning that belongs to the reader. In this book, the author takes the view that climate change is a cyclical, ongoing event and not necessarily the result of human actions. However, the author admits that currently climate change is occurring, and human activities are exacerbating the impact of the changes.

INTRODUCTION

Climate change endangers the health of humans and other living organisms. The environmental consequences of climate change, such as those detailed in this book, including sea level rise; flooding or drought; detrimental effects on drinking water sources; heat waves; degraded air quality; forest fires; vulcanism; earthquakes; and more intense storms such as hurricanes, cyclones, and tornadoes, directly and indirectly affect the health of living organisms, including humans. It is not rocket science to extrapolate climate change drivers manifesting migration (the affected location to the unaffected location)—this is common sense, and is written, well written, and documented. In the attempt to illustrate, describe, explain, and communicate the adverse health effects of climate change, it is interesting to note that we are talking about a double-edged sword.

A double-edged sword?

Yes, absolutely. And actually, on one edge of the sword is a paradox.

How so, you ask?

Okay, the two edges of the sword whereby climate change–driven disease causes migration are presented in the following.

FIRST EDGE OF THE SWORD

Well, to begin with, we can say without too much argument that detrimental climate change events to a certain area, such as contamination, total degradation, or infection by dangerous microbes in the area's water supply, may be the driver to push inhabitants out of the affected area and onto those so-called greener (cleaner and wetter) pastures. When the inhabitants drink polluted water and it makes them sick, and if they survive and are unable to purify the water, they must move on. Now it is true that addressing the effects of climate change on human health is a challenge because both the surrounding environment and the decisions people make affect their health (e.g., they smoke—something, anything—they imbibe alcohol and take illicit drugs).

So exactly what types or kinds of climate change events are likely to harm a lot of people?

One example, a simplified illustration, is that likely climate change–driven increases in the frequency and severity of local heat waves have the potential to harm inhabitants, both human and animal. Of course, the adverse health effects can probably be avoided if pre-planning is effected to the point where preventive measures before the heat wave are enacted, especially for the vulnerable inhabitants—children and the elderly. One thing is certain: there is no room for "feel good" science or other measures when it comes to protecting inhabitants from such climate change events as heat waves. What it takes is to consider a whole host of variables such as socio-economic status, biological susceptibility, the surrounding environment, and cultural competence. Planning for climate change factors that affect the health of inhabitants is not a 1, 2, 3 process; it is complicated. Creating prudent health policies for future climate change events is fraught with uncertainty. How do we predict environmental change and forecast human decisions? Predicting these can be more difficult than predicting the weather. Again, feel-good measures are not the answer; instead, we must employ true science in making our policy decisions.

So on this edge of the double-edged sword related to climate change diseases being a migration driver, the question is: what diseases are we talking about?

Good question. Let's list and discuss the types of climate change diseases that drive inhabitants from their native lands to new ones. The diseases include:[2]

- Asthma, Respiratory Allergies, Airway Diseases
- Cancer
- Cardiovascular Disease and Stroke
- Foodborne Diseases and Nutrition

- Heat-related Morbidity and Mortality
- Human Developmental Effects
- Mental Health and Stress-Related Disorders
- Neurological Diseases and Disorders
- Vector borne and Zoonotic Diseases
- Waterborne Diseases
- Weather-Related Morbidity and Mortality

Before we describe each of these climate change–driven diseases, Sidebar 12.1 is provided to point out the essence, the crux, the core, the heart of air.

SIDEBAR 12.1   AIR

Whether we characterize it as a caress or a light touch against soft skin, as a gentle breeze, a warm wind, blustery gale, as tempest, typhoon, tornado, or hurricane, air is vital.

Air can dry, cool, warm, freshen, ventilate, or disgust.

Air encapsulates us. It surrounds us. We take it in, as we must, with every breath; our bodies thrive on it, and we fail immediately without it. Literally awash with air, on Earth, all life we know of depends on it. It occurs naturally everywhere on Earth—the sky begins where the ground ends.

Air is scientifically unique. The combination of common and rare gases we breathe has made life possible. Air, as with water, is the only chemical compound found naturally that affects most living organisms in a manner of ways.

We associate air with all the good on earth. We cannot imagine life without breathing—we must constantly quench our thirst for air. Air sustains growth. It creates the subtle and blatant movements that provide us with changing weather patterns. But can we really say emphatically, definitively, that air is only good?

No. We cannot. Nothing—absolutely nothing—is safe from air.

Air is odorless, colorless, tasteless. We rarely stop to think about it unless it brings something to us as a reminder. But it covers Earth completely. Nothing can escape air's touch. Nothing.

Air is life—life and air are inseparable.

We sometimes call air the breath of life—a fitting name, especially when you consider that air can be the boon—or bane—of all life, capable in time of sustaining or destroying all life as we define it.

Whether it provides the fundamental source of power in a sailing vessel, pushes a blade of a windmill, a billowy cloud, a dust mote, a feather; whether it lifts a bird soaring on thermals; whether it drifts to us the sweet fragrance of gardenia, lavender, lilac, rose, or a seed to fertile ground; whether it sets water lapping against some distant shore, drives a gritty wind that sculpts mountains to sand, or hammers a horrendous fist that flattens whatever stands in its path—cities, forests, crops—and man—air is essential. Air is life. Air is vital.

Air gives us the blessing of communication. From our first cry to our dying breath, our voices travel on a current of air. Air carries sound. Can we hear a more pleasant sound than wind passing through pine? Can our spines tingle more at the sound generated by wind against an ancient shutter?

Air carries warmth. Air carries cold.

Air is vital.

Our very existence depends on air, but we have created a paradox with our vital line to life. Why would we abuse something so vital—something we need to survive—something without which we cannot live? Why do we foul the very essence of our lives? Why do we insult our environment at a faster pace than we understand and mitigate the consequences? Why? Because air—air—air is everywhere. We've always had enough. Right?

Let us hope that we always will. Let us hope that we are not destroying the very air we breathe. Let us hope that technology will aid us in our efforts to retain the quality of air we need to survive.

We need air as it should be: pure, wholesome, and sweet-smelling, in the perfect mixture of elements we were evolved to inhale, vital to our existence.

Does air care about the greenhouse effect? About ozone depletion? About global warming? No. It cannot.

Should we?

The bottom line: We must not forget that we exist by environmental consent, subject to change without notice.

—F. Spellman (1998)

Asthma, respiratory allergies, and airway diseases—in recent decades, the world has seen a sharp rise in the prevalence (as well as severity) of respiratory diseases, including allergic diseases, asthma, hay fever, and rhinitis (i.e., runny nose, sneezing, nasal stuffiness). A number of experts speculate that the global rise in respiratory diseases such as asthma is indirectly related to climate change (D'Amato and Cecchi, 2008). Note that numerous respiratory allergic diseases are seasonal with climate-sensitive components; climate change may increase not only the incidence of allergic diseases but also intensify them. While it is true that some risk for respiratory disease may be clearly linked to climate change, for some, this link is questionable. It should become a high priority to research the connection between climate change and respiratory diseases. This is the case especially because asthma and respiratory allergic diseases are so prevalent. Having said that, it is important to keep in mind that not all asthmatic episodes are triggered by environmental factors. Other factors include ambient air pollutants, allergens, stress, and a multitude of other environmental variables.

It is not just asthma and other allergic diseases that are a concern for people exposed to climate change. Climate change has the potential to impact airway diseases by increasing ground-level ozone and possibly fine particle concentrations. Keep in mind that fine, respirable-size particulates are in the size range that permits them to penetrate deep into the lungs upon inhalation.

CANCER

If there is any word in any language that brings about instant diligence or instant dread, fear, and concern, especially if it is with regard to us, it is the word cancer. Technically, cancer refers to a group of diseases in which abnormal cells divide without control and are able to invade other tissues. At the present time, we know of at least 100 different types of cancer, and they are generally referred to by the organ or type of cell in which they arise (e.g., breast, prostate, colon, lung, liver). Cancer is the second leading cause of death in the United States after heart disease, killing more than half a million people every year (National Cancer Institute, 2009).

Climate change and cancer go hand in hand—both directly and indirectly. Climate change results in higher ambient temperatures that may increase the transfer of volatile and semi-volatile compounds from water and wastewater into the atmosphere and alter the distribution of contaminants to places more distant from the sources, changing subsequent human exposures. Also, climate change is expected to increase heavy precipitation and flooding events, which could increase the chance of toxic contamination tasks from storage facilities or runoff into water from land containing toxic pollutants. Note that very little is known about how such transfers will affect people's exposure to these chemicals—some are known carcinogens—and its ultimate impact on cancer (Bates et al., 2008).

Although the precise mechanisms of cancer in humans and animals are not completely understood for all cancers, factors in cancer development include pathogens, environmental contaminants, age, and genetics. Because of the challenges of understanding the causes of cancer, the links between climate change and cancer are a mixture of fact and supposition or guesswork.

With regard to the direct impact of climate change on causing cancer, it is possible that an increase in exposure to toxic chemicals (that are known or suspected to cause cancer) following heavy rainfall, and by increased volatilization of chemicals under increased temperatures, may enhance the chance of causing cancer. Note that heavy rainfall and/or flooding may cause an increase in leaching of toxic chemicals and heavy metals from storage sites and increased contamination of water with runoff containing "forever chemicals" (persistent chemicals) that are already in the environment. Aquatic animals, including mammals, also may suffer direct effects of cancer linked to sustained or chronic exposure to chemical contaminants in the marine environment and thereby serve as indicators of similar risks to humans (McAloose and Newton, 2009).

Another direct effect of climate change, depletion of stratospheric ozone, results in increased ultraviolet radiation exposure.

Does the ozone hole portend disaster right around the corner?

Maybe you've seen the numerous headlines: "1997 Was the Warmest Year on Record," "Scientists Discover Ozone Hole Is Larger Than Ever."

We have already discussed the global climate problem, but let's take a look at the stratospheric ozone problem.

Ozone is formed in the stratosphere by radiation from the sun and helps to shield life on Earth from some of the sun's potentially destructive ultraviolet radiation.

In the early 1970s, scientists suspected that the ozone layer was being depleted. By the 1980s, it became clear that the ozone shield was indeed thinning in some places, and at times, even has a seasonal hole in it, notably over Antarctica. The exact causes and actual extent of the depletion are not yet fully known, but most scientists believe that various chemicals in the air are responsible.

Most scientists identify the family of chlorine-based compounds, most notably chlorofluorocarbons and chlorinated solvents (carbon tetrachloride and methyl chloroform) as the primary culprits involved in ozone depletion. In 1974, Molina and Rowland hypothesized that CFCs (containing chlorine) were responsible for ozone depletion. They pointed out that chlorine molecules are highly active and readily and continually break apart the three-atom ozone into the two-atom form of oxygen generally found close to Earth in the lower atmosphere.

The Interdepartmental Committee for Atmospheric Science (1975) estimates that a 5 percent reduction in ozone could result in nearly a 10 percent increase in cancer. This already frightening scenario was made even more frightening by 1987 when evidence showed that CFCs destroy ozone in the stratosphere above Antarctica every spring. The ozone hole had become larger, with more than half of the total ozone column wiped out and essentially all ozone disappearing from some regions of the stratosphere (Davis and Cornwell, 1991).

In 1988, Zurer reported that on a worldwide basis, the ozone layer had shrunk approximately 2.5 percent in the preceding decade. This obvious thinning of the ozone layer, with its increased chances of skin cancer and cataracts, is also implicated in suppression of the human immune system and damage to other animals and plants, especially aquatic life and soybean crops. The urgency of the problem spurred the 1987 signing of the Montreal Protocol by 24 countries, which required signatory countries to reduce their consumption of CFCs by 20 percent by 1993 and by 50 percent by 1998, marking a significant achievement in solving a global environmental problem.

The Clean Air Act of 1990 borrowed from EPA requirements already on the books in other regulations and mandated phase-out of the production of substances that deplete the ozone layer. Under these provisions, the EPA was required to list all regulated substances along with their ozone-depletion potential, atmospheric lifetime, and global warming potentials.

We know that UV radiation exposure increases the risk of skin cancers and cataracts (Tucker, 2009). Note that the incidence of typically nonlethal basic cell and squamous cell skin cancers is directly correlated to the amount of exposure to UV radiation. Several other variables compound the effect, including temperature and exposure to other compounds that can amplify the carcinogenic potential of UV radiation (Burke and Wei, 2009). UV exposure increases with rising temperatures (such as those that occur at night versus day and in summer versus winter). If increases in average or peak temperatures occur as a result of climate change, an increase in the incidence of non-melanoma skin cancer may occur (van der Leun and Place, 2008). Previous research has shown that increased UV radiation exposure combined with certain polycyclic aromatic hydrocarbons (PAHs) can enhance the phototoxicity of these compounds and damage DNA (Dong et al., 2000). Increased UV radiation also could impact the human immune system and alter the body's ability to remove the earliest mutant cells that begin the cancer process, although it is unclear whether these changes would be beneficial or detrimental.

CARDIOVASCULAR DISEASE AND STROKE

There is evidence of climate sensitivity for cardiovascular diseases (i.e., for chest pain, acute coronary syndrome, stroke, and variations in cardiac dysrhythmias), with both cold and extreme heat affecting the incidence of hospital admissions (Bassil et al., 2009). Extreme heat generated by weather serves as a stressor in individuals with pre-existing cardiovascular disease and can directly precipitate exacerbations (Fouillet et al., 2006). In addition, there is evidence that heat amplifies the adverse impacts of ozone and particulates on cardiovascular disease.

FOODBORNE DISEASES AND NUTRITION

Humans and other living organisms take in food and use it for growth and nourishment. The sum of these processes is nutrition (a.k.a. sustenance). Together with clean air, water, and shelter, nutritious food is a basic necessity of life. Without sufficient calories and an adequate mixture of macronutrients (calories, fats, proteins, carbohydrates), micronutrients (vitamins, minerals), and other bioactive components of food, death can result. Extreme weather events and changes in temperature and precipitation can directly damage crops or other food supplies, as well as interrupting supply support systems such as transportation methods that distribute food supplies. Although this usually is attributed to seasonal occurrences, it can become a chronic problem under changing climate conditions. Note that indirectly there is the potential for harm from undernutrition or even famine resulting from damage to agricultural crops and related trade, economic, and social instability; diversion of staple crops for use in biofuels (corn for ethanol or other biofuels); changes in agricultural practices, including those intended to mitigate or adapt to climate change; compromised ability to grow crops due to changing environmental conditions and water availability; and diminished availability nutritional quality and value of protein from fisheries, aquaculture, and other marine-based foods.

DID YOU KNOW?

Currently, two methods are commonly used for estimating climatic impacts on agriculture. The two methods used to estimate the effect of climate on crop production are: (1) structural modeling of crop and farmer response, combining the agronomic response of plants with economic/management decisions of farmers, and (2) spatial analogue models that exploit observed differences in agricultural production and climate among regions. These approaches are complementary. Reconciling differences in results between these methods enables better understanding of agricultural adjustment to climate change. Uncertainty will necessarily remain because of the nature of climate and agricultural production (Schimmelppfennig et al., 2021).

Although food is the primary source of essential nutrients, food can also be a source of exposure for foodborne illness. These illnesses result from ingesting food that is spoiled or contaminated with microbes, chemical residues such as pesticides, biotoxins, and other toxic substances. Seafood contaminated with metals, biotoxins, toxicants, or pathogens; crops burdened with chemical pesticide residues or microbes; extreme shortages of staple foods; and malnutrition are possible effects of climate change on the production, quality, and availability of food (IPCC, 2007). The potential effects of climate change on foodborne illness, nutrition, and security, on a global scale, are huge in terms of numbers of people likely to be affected and consequent human suffering. Some of these effects are already being felt in the wake of extreme weather events such as droughts, flooding, and hurricanes, and as such present a fairly immediate concern (Schimmelppfennig et al., 2021).

SIDEBAR 12.2   *V. PARAHAEMOLYTICUS, V. VULNIFICUS, V. CHOLERA*

Climate change may impact rates of foodborne gastroenteritis through increased temperatures. Several species of *Vibrio (V)*, naturally occurring marine bacteria, are sensitive to naturally occurring bacteria and changes in ocean temperature. *Vibrio parahaemolyticus* infects oysters and is the leading cause of *Vibrio*-associated gastroenteritis in the United States. Note that an outbreak in Alaska in 2004 has been linked to higher-than-normal ocean temperatures (McLaughlin et al., 2005). Other studies show a predictive relationship between sea surface temperature and *V. vulnificus* and *V. cholera*. Climate-driven changes in ocean temperature and coastal water quality are expected to increase the geographic range of these bacteria and could be used to predict outbreaks. Increased temperatures also affect other foodborne illnesses, including camplyobacteriosis and salmonellosis.

As a sidelight, so to speak, I want to point out that this author was hired to conduct an investigative case for the U.S. Department of Justice in 2007. I looked into the case aboard a U.S. Naval ship homeported in Japan. The case involved a lawsuit filed by a civilian contractor working on equipment aboard the ship. In the process of troubleshooting equipment failure, the contractor was contaminated with a small quantity of fluid that leaked from a ruptured pipe, down a drain, and through a crack in the deck above the worker's head. Two weeks later, while in the United States, the contractor's right arm swelled to double its size and turned black, accompanied with severe pain. In the hospital the contractor was diagnosed as having been infected with *V. vulnificus* that had manifested into methicillin-resistant *Staphylococcus aureus* (MRSA)—the flesh-eating bacteria. The contractor's arm was saved, and he fully recovered, but he filed a lawsuit against the Navy and federal government—remember, this is the era of "let me sue you before you sue me."

During my investigation I was able to obtain a sample of the fluid that had leaked onto the contractor; the ship's crew had saved a sample of the fluid and properly labeled and stored it. When I used the ship's laboratory to analyze the fluid, I found that it was just plain old water, with no accompanying contaminates. Thus, I determined that the fluid could not have given the contractor MRSA.

So what caused the contractor's infection?

After a few days of research and investigation aboard ship and on shore outside the base in Japan, in the local town, I found the cause of the contractor's MRSA. A couple of the Navy personnel who had collaborated with the contractor informed me that he had eaten dinner ashore with them on more than one occasion, and that his favorite meals included large portions of shellfish. On hearing this I immediately shifted suspicion of the causal factor from anything aboard the ship and instead focused on news that due to warmer water temperatures in this area, some of the shellfish might be contaminated with pathogens, including *V. vulnificus*. Fortunately for the contractor, medical professionals found that he had indeed been infected with *V. vulnificus* and that the probable source was the shellfish the contractor had ingested for two weeks while in Japan. I also verified that the Naval ship did not serve any type of shellfish on board during the contractor's stay.

Case solved.

But this was only the first case involving a contractor contracting MRSA after having worked aboard a U.S. Naval ship. I accepted assignment from U.S. Justice Department to investigate this particular case in Bremerton, Washington, and later at Submarine Base Bangor, also in Washington State a few miles north of Bremerton. In this case an engineer was doing alignment work in the bilges of the vessel. During his work there was a raw sewage overflow event that occurred a deck above the bilge area, causing the contractor to come in contact with the raw sewage. A couple of weeks later, after he had returned home to the U.S. east coast, he became ill and hospitalized and was diagnosed with a MRSA infection that was beginning to do damage to his body. Fortunately, Naval doctors quickly determined that he did indeed have MRSA. The medical staff was able to act quickly and use the correct antibiotics to arrest the infection and save the engineer's infected body parts and possibly his life. Meanwhile, the victim had hired an attorney and filed a lawsuit against the Navy for contaminating him with raw sewage that in turn led to his MRSA infection.

Because I was an expert in water/wastewater treatment with dozens of published handbooks in the operation of water and wastewater treatment plants, conveyance systems, and hydraulics and also a microbiology book for water/wastewater operators, I was thought to be an expert on MRSA-type pathogens being residents of wastewater (sewage) systems. Truth be told, I was not aware of MRSA pathogens residing in wastewater, because I had not studied that facet of the putrid mix. So when assigned this case, I ran a quick, non-scientific sampling and analysis of wastewater at five different wastewater treatment plants in Virginia. I sampled raw influent and the various unit processes downstream in the system that ended with samples of chlorine-treated wastewater and also sampled and tested the wastestream at the outfall.

During my non-scientific testing and analysis, using my personal protocols, I was surprised to find that MRSA-causing pathogens were passengers within the raw sewage influent. As the wastewater passed from one unit process to the next, I found it harder to detect MRSA-causing pathogens, and by the time the wastewater was treated with chlorine, I found no evidence of MRSA-causing pathogens. Later, shifting from secondary treatment plants to a tertiary treatment plant, again I found no MRSA-causing pathogens.

The point is, my theory that MRSA-causing pathogens were probably not present in wastewater because of its hostile environment for certain pathogens proved wrong. And since the infected engineer was exposed to untreated, raw sewage, there was a possibility that he did receive his contamination from the raw wastewater spill on the Naval ship.

However, I was not totally convinced that his infection was the result of the wastewater spill on that Naval ship. It did not feel right to me. So I did additional investigation. I visited his home on the east coast and found out that his wife and the couple's two children also had MRSA infections but had been properly treated and were fully recovered. And, surprisingly, the wife and children were infected with MRSA while the husband was away working on the Naval ship. The husband did not contract MRSA, or it did not manifest itself, until he had a conjugal visit with his wife.

Note that the wife and husband were separated but still friends and met once in a while for intimacy, and so forth, and so on. Anyway, I did not have to call upon Sherlock Holmes-type deductive reasoning to conclude the causal factors of the family's MRSA infection. No. Instead the wife was very open, straightforward, and honest with me when I asked her questions about the infections. First, she voluntarily provided a complete explanation of her contracting MRSA as far as she understood it. My main questions dealt with whether she had any idea of where she contracted the disease. She said at first, she had no clue, but after a couple of her patients had been diagnosed with the same malady and were receiving treatment, she assumed she contracted the disease from them and passed it on to her family. So I asked what type of doctor she was because she had mentioned her patients. She replied that she was a contract masseuse, not a medical person, and that her patients were actually customers.

Needless to say, her stating that she was a masseuse got my immediate attention because she had stated that two of her customers were also infected with MRSA. And while she explained her experience with the diseased customers, I glanced around the interior of her small house and noticed at the end of a short hallway stood a neatly stacked pile of linen sheets and alongside the washing machine was a pile of crumpled-up linens. And I continued to listen while I looked around. Finally, we got around to talking about her large stacks of clean and used linens. She explained that they were part of her business and that she brought the used linens home and washed them for reuse.

This information was important to me. The linens were used on her customers when she was massaging them, and to me this meant that the two customers who were contaminated probably contaminated the linens, and when she brought them home, her family members were exposed to the contaminated linens and became infected.

I made my report to the Justice Department Assistant Attorney, and within a week the plaintiff dismissed the case.

Case closed.

Note: I provided the preceding sidebar information as information only and to point out that danger is everywhere, and conditions related to climate change can only exacerbate exposure to diseases that are all around us. I provided the information in plain English for understanding and avoided going into detail about septicemia, the severity of wound infections, detailed gastroenteritis generated by exposure to MRSA-causing diseases—this was done to ensure understanding.

Getting back to foodborne illness and its connection to climate, note that the U.S Climate Change Science Program (CCSP) reported a likely increase in the spread of several foodborne pathogens due to climate change, depending on the pathogen's survival, persistence, habitat range, and transmission in the changing environment (Gamble and Ebi, 2008). Crop pests such as aphids, locusts, and whiteflies, as well as the spread of the mold *Aspergillus flavus* that produces aflatoxin, a substance that may contribute to the development of liver cancer in people who eat contaminated corn and nuts are encouraged during drought periods. Agronomists are concerned that climate-change–based increases in a variety of blasts, rusts, blights, and rots will further devastate already stressed crops and thereby exacerbate malnutrition, poverty, and the need for human migration. The problem is that the spread of agricultural pests and weeds may lead to the need for greater use of some toxic chemical herbicides, fungicides, and insecticides (Gregory et al., 2009), resulting in potential intermediate hazards to farm workers and their families (Lynch et al., 2009), as well as longer-term hazards to consumers, particularly children (Eskenazi et al., 2008).

Extreme weather events due to climate change, particularly flooding, drought, and wildfires, can impact the safety of agricultural crops and fisheries by exposure to and contamination from metals, chemicals, and other toxicants (Ebi et al., 2008). Contaminant and pathogen pathways will be altered by global climate changes in ocean currents and water mass distribution, along with changes in Arctic ice cover, length of melt seas, hydrology, and precipitation patterns (Ebi et al., 2008). Contaminants include a wide range of chemicals and metals such as polychlorinated biphenols (PCBs), polycyclic aromatic hydrocarbon, mercury, and cadmium; pharmaceuticals such as synthetic hormones, statins, and antibiotics; personal care products; widely used industrial chemicals such as fire retardants, stain repellants, non-stick coatings (per- and polyfluoroalkyl substance, PFAS); and pesticides and herbicides for agricultural use and vector control for public health protection.

The CCSP noted the strong association between sea surface temperature and proliferation of many *Vibrio* bacteria species (recall Sidebar 12.2) that occur naturally in the environment and suggested that rising temperatures would likely lead to increased occurrence of illness associated with *Vibrio* bacteria in the United States, especially seafoodborne disease associated with *V. vulnificus* and *V. parahaemolyticus* (Ebi et al., 2008). More virulent strains of existing pathogens and changes in their distribution or the emergence of new pathogens could manifest and multiply with rising temperatures and impacts on other environmental parameters such as ocean acidification (Smolinski et al., 2003). Those highly dependent on marine-based diets for subsistence, such as the natives in Alaska, may be exposed to increased risks from animalborne disease pathogens (Sokurenko et al., 2006). Climate change–induced acidity of water may alter environmental conditions, leading to greater proliferation of microbes of public health concern. This is a considerable concern in molluscan shellfish because ocean acidification may affect formation of their carbonate shells and their immune response, making them more vulnerable to microbial infection.

SIDEBAR 12.3   COVID-19/PFAS—AND THE PIPES WILL TELL

In 2020, while I was doing research for my book, *Fundamentals of Wastewater-Based Epidemiology: Biosurveillance of Bacteria, Protozoa, COVID-19 and Other Viruses*, I used modern epidemiological techniques to identify bacteria, fungi, COVID-19, and other viruses as a means of measurement. Within this text, a "means of measurement" is defined as using wastewater to identify the predominance of pathogenic organisms specific to sources (populations) in an area, region, county, and city.

Wastewater-based epidemiology (WBE), as used in this book, is another important term to define. WBE is sometimes labeled as, or discussed as,

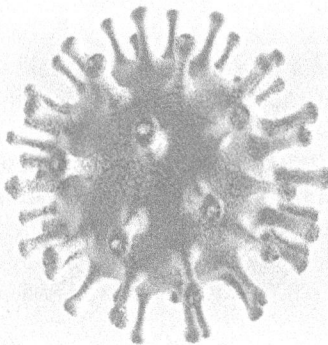

FIGURE 12.1   COVID-19.

Source: Illustration by Kat Welsh-Ware and F. Spellman.

wastewater-based surveillance, sewage biological information mining, or sewage chemical information mining; it is WBE surveillance used to mine for biological information in wastewater that is the focus herein. In light of this, WBE information mining is the technique used for the consumption of or exposure to pathogens in a population. This is accomplished by measuring (the key word is "measuring") biomarkers (i.e., biological entities) in wastewater produced by people contributing human wastes to wastewater treatment plants.

Well, there are some who might believe that WBE is an emerging technology—that it must be new and developing. Actually, WBE has been used for years in the measurement of illicit drug use (and $100 bills) in populations in various locations. Where does the money come from? Well, when someone is a drug dealer with drugs and dollars in their possession, and then the law knocks on their door with a search warrant, it is amazing to many people what is suddenly flushed down the toilet, so to speak. Raw wastewater influent into a wastewater treatment plant is often and commonly used to measure illicit drug use and also measure the consumption of an assortment of personal care products, compounds, pharmaceuticals, nicotine, opioids, artificial sweeteners, alcohol, caffeine, and others. WBE has been modified to measure the pack of pathogens such as SARS-CoV-2 in a particular area or region—note that this is a measure of a population as a whole. It should be pointed out that WBE is an interdisciplinary effort that draws on input from specialists such as wastewater treatment plant professionals, chemists, environmental scientists, laboratory technicians, and epidemiologists.

WBE of pathogenic organisms has the possibility to inform on the presence of a disease outbreak when or where it is not assumed or expected. At the present time we are concerned with the COVID-19 pandemic. Wastewater-based epidemiology is being employed to test for the presence of SARS-CoV-2 in wastewater. The point is wastewater can be tested for signatures of viruses, including SARS-CoV-2 excreted via feces (Medema et al., 2020; Okoh et al., 2010; Gundy et al., 2008).

In August 2020, the author conducted an informal survey of wastewater treatment plants in the United States and found more than 100 sanitation districts and publicly owned treatment works that were conducting wastewater surveillance of SARS-CoV-2 as a potentially important source of information on the prevalence and chronological trends of COVID-19 in communities. For example, one of the globe's premier wastewater treatment operations, Hampton Roads Sanitation District in southeastern Virginia, is actively pursuing collection, study, research, and testing of wastewater influent in the major treatment plants.

With regard to HRSD and its realization that sewage can suggest a wider spread of COVID or no spread at all, the environmental scientists, microbiologists, technical services specialists, and chemists in the organization are actively taking samples for COVID-19 testing. The fact is wastewater has proven to be an early indicator of the next outbreak. It is interesting to note that the 2020 summer increase in the Hampton Roads region of southeastern

Virginia was being detected not through nose swabs or other medical testing but instead in the wastewater (sewer) pipes—"And what's in the pipes will tell."

The professionals saw an increase in genetic material from the virus surge in wastewater being produced by hundreds of thousands of Hampton Roads residents when samples were analyzed from the district's treatment plants. As the district's data were plotted over the health department's data, pretty close similarities were seen, although there was a lag time of one week from the time signals were apparent in the wastewater analyzed and the time clinical data started to change.

The process employed by HRSD on sampling, testing, and analyzing raw sewage samples is wastewater-based epidemiology. The advantage of using WBE is detection of the virus's presence soon after people flush their toilets. It is a heck of a lot quicker than waiting for people to be tested and get their results back. While SARS-CoV-2—its official scientific name—is primarily thought of as a respiratory pathogenic virus, scientists have found that it can affect the digestive system, and the virus's genetic makeup can be detected in stool specimens even before symptoms manifest themselves and are observable or felt. Tracking the coronavirus through wastewater can be an early indicator of where the next outbreak could happen. In other words, using WBE-provided information can allow decisions to be made before things get worse. And that is the beauty, the goal, the objective, and the purpose of employing WBE.

One of the surprises I had was when I looked at various samples of raw wastewater and their accompanying laboratory analyses, I noticed that the samples also contained personal care products, hormones, and other assorted ingredients—these were not the surprise—instead, minute amounts of the PFAS (so-called "forever chemicals") were also present, and that was a surprise. I measured these in quantities of parts per trillion (ppt)—very small quantities, for sure, but present and extremely persistent, extremely.

So what are PFAS?

Good question.

Answer: PFAS (a.k.a. forever chemicals) are human-made chemicals used in a wide range of consumer and industrial products. PFAS do not easily break down, and some types have been shown to accumulate in the environment and in our bodies. The point is we need to be aware of PFAS and conduct further research on their impact on the environment and living organisms.

HEAT-RELATED MORBIDITY AND MORTALITY

In 2001, Houghton et al., working with the IPCC, published their *Climate Change 2001: The Scientific Basis*, where they state that as a result of anthropogenic (human-caused) climate change, global mean temperatures are rising and are expected to continue to increase regardless of reducing greenhouse emissions. I have stated right up front in this book that it is my belief that climate change is cyclical, and the Earth has experienced several episodes of cold-to-warm and warm-to-cold events throughout its history. I also have stated that it is my belief that even if humans are not the causal factor of climate change, we certainly are exacerbating the problem via our polluting activities—global climate change is occurring. Note that IPCC refers to climate change as any change to climate over time, whether due to natural variability or as a result of human activity—this opinion is one I fully agree with. More, I am in full agreement with the Houghton group's statement that climate change will continue regardless of our progress in reducing greenhouse gas emissions—we have crossed the bar, so to speak. In 2007, the IPCC stated that global average temperatures are projected to increase between 1.8 and 4.0°C by the end of this century (IPCC, 2007). Climate change is expected to raise overall temperature distribution and to contribute to an increase in the frequency of extreme heat events, or heat waves (Meehl and Tebaldi, 2004). Temperature (particularly temperature extremes) is associated with a wide range of health impacts and, in some locations, is a driver of migration. For example, consider northern Africa and the Middle East, where intensification of heatwaves could be a significant driver of migration from the region (Workman, 2021).

The health outcomes of prolonged heat exposure include heat exhaustion, heat cramps, heat stroke, and death (Spellman and Stoudt, 2013). In the United States, extreme heat events cause more deaths annually than all other extreme weather events combined. In the United States an average of almost 700 people succumb to heat-related death per year (Spellman and Stoudt, 2013). Prolonged exposure to heat may result in additional illness and death by exacerbating preexisting chronic conditions such as various respiratory, cerebral, and cardiovascular diseases (Kovats and Hajat, 2008), as well as increasing risk for patients taking psychotropic drug treatment for mental disorders (Davido et al., 2006), due to the body's impaired ability to regulate temperature.

DID YOU KNOW?

During the August 2003 heat wave in France, Davido et al. (2006) conducted a study to determine the risk factors for short-term mortality in the victims of the heat wave from among patients evaluated in the emergency department. The theory was that age, temperature, and some long-term therapies and pre-existing pathologies were factors associated with short-term mortality. Their findings showed that of 841 patients attending the emergency department in the study period, 165 were included in the study, of which most were elderly women. Thirty-one (18.8 percent) died within one month. Factors associated with short-term mortality were a greater degree of dependent living; more severe clinical condition on admission (higher temperature and heart rate, lower blood pressure, hypoxia, and altered mental status); higher values of blood glucose, troponin, and white blood cell count; lower values of serum protein and prothrombin levels; pre-existing ischemic cardiomyopathy; pneumonia as associated infection; and previous psychotropic treatment. The total number of survivors at 1 year was 91.

HUMAN DEVELOPMENTAL EFFECTS

Some birth defects have been steadily increasing over the last couple of decades; for example, the rates of congenital heart defects have doubled (Correa-Villasenor, 2003). This suggests a possible link to environmental effects, although other explanations are possible, such as better reporting. We simply do not know what we do not know about the increased occurrence of birth defects.

Anyway, what we do know for sure is that most humans develop in a predictable manner, growing from a fertilized egg to fetus, newborn, toddler, child, adolescent, and in adult in a way that is fairly well known. It is also well known that the environment can be a potent modifier of normal development and behavior. Not all environmental effects on development are mainstream, so to speak, that is, visible and unmistakable. Some are subtle, like impact on IQ from exposure to lead—this has been an ongoing problem with children living in older homes or poverty who chew on paint containing lead on windowsills and other locations. Moreover, changes in onset of puberty can occur from exposure to endocrine-disrupting chemical compounds.

With regard to endocrine-disrupting chemical compounds, according to USEPA (1998), it has been suggested that humans and domestic and wildlife species have suffered adverse health consequences resulting from exposure to environmental chemicals that interact with the endocrine system. However, considerable uncertainty exists regarding the relationship(s) between adverse health outcomes and exposure to environmental contaminants. Collectively, chemicals with the potential to interfere with the function of endocrine systems are called endocrine-disrupting chemicals (EDCs). EDCs have been broadly defined as exogenous agents that interfere with the production, release, transport, metabolism, binding, action, or elimination of the natural hormones responsible for the maintenance of homeostasis and the regulation of developmental processes.

It is important to point out that EDCs, along with, in a broad sense, pharmaceutical and personal care products (PPCPs), fall into the category of "emerging contaminants." Emerging contaminants can fall into a wide range of groups defined by their effects, uses, or key chemical or microbiological characteristics. These compounds are found in the environment, often as a result of human activities (Spellman, 2014).

Other possible environmental-related birth defects (a slight association) include cleft lip due to dioxin-like compounds (Pradat et al., 2003). Another problem is fetal loss through exposure-related spontaneous abortion (Bukowski, 2001). According to the Centers for Disease Control and Prevention, about 3 percent of all children born in the United States have a birth defect, some of which can be attributed to environmental causes. Birth defects are a leading cause of death in children, accounting for almost 20 percent of all infant deaths. CDC (2022) reports that about 1 in every 33 babies is born with a birth defect. CDC also acknowledged that not all birth defects can be prevented.

Environmental exposures during vulnerable periods during human development, including preconception (gametogenesis), preimplantation, the fetal period, and early childhood can lead to functional deficits and developmental changes through several mechanisms, including genetic mutations and epigenetic change (epigenetics is the study of the behavior genes and how gene expression can be altered by the environment without changes in DNA). Note that some chemicals damage DNA directly, causing mutations in gametes or the developing fetus that can lead to later disease or conditions that increase disease risks such as obesity (Wadhwa et al., 2009).

When addressing or speaking about environmental changes and developmental origins of health and disease and the epigenetic mechanisms involved, it is important to take a look at what is known as "Barker's hypothesis" (a.k.a. fetal origins hypothesis), which emerged 25 years ago from epidemiological studies of infant mortality. David J. Barker, a British physician and epidemiologist, has argued that inadequate nutrition *in utero* "programs" the fetus to have metabolic characteristics that can lead to future disease (1992). For example, Barker argues that individuals starved *in vitro* are more likely to become overweight as adults and that they are more likely to suffer from diseases associated with obesity, including cardiovascular problems and diabetes (Almond and Currie, 2014).

The key point (at least for this book) about Barker's hypothesis is the connection between development and environment because problems with environment, especially, can be a driver of migration to those so-called greener pastures. Actually, Barker's hypothesis combines several key ideas in relation to the fetal origins hypothesis. First, the effects of fetal conditions are persistent. Second, there is a latency period whereby the health effects can remain hidden for years—a good example is heart disease, which does not emerge as a problem until middle age. Third, the hypothesized effects reflect a specific biological mechanism, fetal "programming," possibly through effects of the environment on the epigenome, which are just beginning to be understood—this is another one of those cases where we do not know what we do not know, but we are working on it. Petronis (2010) points out that the epigenome can be conceived of as a series of switches that cause various parts of the genome to be expressed—or not. The period while the fetus is *in vitro* may be particularly important for setting these switches.

The bottom line: Climate change effects on food availability and nutritional content could have a marked, multigenerational effect on human development (Spellman, 2014) and could provide the driver for migration to greener pastures where food is readily available, with the accompanying nutritional benefits.

SIDEBAR 12.4   DDT AND HUMAN PREGNANCY

We are quickly approaching a timeline in which we will enter into the fifth generation of people exposed to toxic chemicals from before conception to adulthood. In a few cases, we have identified the hazards of certain chemicals and their compounds and have implemented restrictions. One well-known chemical compound that comes to mind with regard to its environmental harm and subsequent banning is dichloro-diphenyl-trichloroethane (DDT). Let's take a look at DDT.

The insecticide DDT (a.k.a. a legacy pesticide) was first produced in the 1920s and was developed as the first of the modern synthetic insecticides in the 1940s. It was extensively used between 1945 and 1965 with great effect to control and eradicate insects that were responsible for malaria, typhus, and other insectborne human diseases among both military and civilian populations and for insect control in crop and livestock production, institutions, homes, and gardens. DDT was an excellent insecticide because it was very effective at killing a wide variety of insects at low levels. DDT's quick success as a pesticide and broad use in the United States and other countries led to the development of resistance by many insect pest species. Moreover, the chemical properties that made this a good pesticide also made it persist in the environment for a long time.

This persistence led to accumulation of the pesticide in non-target species, especially raptorial birds (e.g., falcons). Due to the properties of DDT, the concentration of DDT in birds could be much higher than concentrations in insects or soil. Birds at the top of the food chain (e.g., pelicans, falcons, eagles, and grebes) had the highest concentrations of DDT. Although the amount of DDT did not kill the birds, it interfered with calcium metabolism, which led to thin eggshells. As a result, eggs would crack during development, allowing bacteria to enter, which killed the developing embryos. This had a great impact on the population levels of these birds. Peregrine falcons and brown pelicans were placed on the endangered species list in the United States, partially due to declining reproductive success of the birds from DDT exposure.

Rachel Carson, that unequaled environmental journalist of profound vision and insight, published *Silent Spring* in 1962, helping to draw public attention to this problem and to the need for better pesticide controls. This was the very beginning of the environmental movement in the United States and is an excellent example of reporting by someone affiliated with the media that identified a problem and warned of many similar problems that could occur unless restrictions were put in place related to chemical pesticide use. Partially as a result of Carson's flagship book, scientists documented the link between DDT and eggshell thinning. This led the U.S. Department of Agriculture (the federal agency with responsibility of regulating pesticides before the formation of the U.S. Environmental Protection Agency in 1970) to begin regulatory actions in the later 1950s and 1960s to prohibit many of DDT's uses because of mounting evidence of the pesticide's declining benefits and environmental and toxicological effects.

In 1972, EPA issued a cancellation order for DDT based on adverse environmental effects of its use, such as those to wildlife (the known effects on other species, such as raptors), as well as DDT's potential human health risks. Since then, studies have continued, and a causal relationship between DDT exposure and reproductive effects is suspected. Today, DDT is classified as a probable human carcinogen by U.S. and international authorities. This classification is based on animal studies in which some animals developed liver tumors.

DDT is known to be very persistent in the environment, will accumulate in fatty tissues, and can travel long distances in the upper atmosphere. Since the use of DDT was discontinued in the United States, its concentration in the environment and animals has decreased, but because of its persistence, residues of concern from historical use remain. Moreover, DDT is still used in developing countries because it is inexpensive and highly effective. Other alternatives are too expensive for these other countries to use (EPA, 2013).

DID YOU KNOW?

EPA (2013) reports that eagles and other birds are thriving after the 1972 DDT ban. Populations of relatively rare birds such as the bald eagle, the brown pelican, the osprey, and the peregrine falcon are now increasing and becoming more visible in many U.S. fish and wildlife national refuges.

DDT and many of the chemicals that are used to control pests and improve crop yields can affect human development and function as a driver for those concerned to seek greener pastures without the chemicals. Climate change alters rainfall and temperature in various parts of the globe. In some places, climate will, and has, led to changes in agricultural practices for crop yields that could increase pesticide use and thereby increase human and animal exposures. Remember that migration of humans and animals for whatever reason can and usually does apply to insects also. Thus, changes in the range of mosquitoes and other pests that can carry disease also may lead to an increase in the use of DDT and other pesticides. Although malaria is rare in the United States, our open borders and the current rush of migration activities (legal and illegal) have increased the number of cases (imported or indigenous) in some locations (WHO, 2008). As pointed out earlier, DDT is highly efficient for the control of mosquitoes that are capable of transmitting malaria to humans. Even though DDT has been withdrawn from use in the United States, it is still used as a desperate, risky expedient to control mosquitoes in malaria-endemic areas around the world. DDT and its principal metabolite, DDE, are persistent in the environment and in humans. Research indicates that women whose mothers had high DDT levels in their blood when they were *in utero* have shortened menstrual cycles and a reduced chance of becoming pregnant.

MENTAL HEALTH AND STRESS-RELATED DISORDERS

Living in a region, any region, when a climate-driven environmental disaster strikes may motivate many of the inhabitants to migrate to those so-called greener pastures (i.e., move on to a better and more promising location). Note that health of populations depends on the availability of clean air, safe drinking water, food, and sanitation (basic hygiene); exposure to pathogens, toxins, and environmental hazards; and numerous genetic, behavioral, and social factors (Barrett et al., 2014). When an inhabitant is faced with such a life-changing situation and life-altering decision, it is not a far reach to expect the migrant to experience some type or form of mental health impact. Moreover, because of the migrant's exposure to climate-driven environmental disaster, it is logical to expect the migrant to experience a great deal of stress. Stress— human beings have moved from place to place since time immemorial. The motives for and the extent of these migrations put particular stress on individuals and their families. Thus, truth be told, it is not a stretch to state that mental health disorders comprise a wide class of illnesses, from social anxiety to paranoia, to severe disease, including depression and suicidal ideation. The problem is that many mental health disorders can also lead to other chronic diseases and even death. Stress-related disorders derive from abnormal responses to acute or chronic anxiety and include diseases such as obsessive-compulsive disorder and post-traumatic stress disorder.

Note that psychological impacts of climate change range from mild stress responses to chronic stress or other mental health disorders. These are generally indirect and have only recently been considered among the collection of health impacts of climate change (Fritze et al., 2008). Mental health concerns are among the most potentially devastating effects in terms of human suffering and the most difficult to quantify and address. Truth be told, the change in climate has posed unparalleled threats to the health of human beings through impacts on water and food security, as well as heat waves, violent storms, droughts, and rising sea levels, resulting in contagious disease (Barrett et al., 2014).

It is important to point out that climate change has the potential to create sustained natural and humanitarian disasters beyond the scale of those we are experiencing today, which may exceed the capacity of the affected public health systems to cope with common demands (The CNA Corporation, 2007). This was demonstrated in the United States in 2005 by the devastating impact of Hurricane Katrina.

The bottom line on mental health and migration is that migration can lead to increase in culture shock and can have different impacts on different people; culture conflict can produce high rates of attempted suicide, and social support may function as a protector.

NEUROLOGICAL DISEASES AND DISORDERS

Evolving studies suggest that contact with (exposure to) a number of agents whose environmental existence may increase with climate change may have effects on neurological development and functioning. Onset of diseases such as Alzheimer's disease (AD) and Parkinson's disease (PD) is occurring at earlier ages across the population, possibly related to environmental factors related to climate change, or at least exacerbated by exposure to climate change. Exposure to pesticides and herbicides, for example, during specific development windows, in combination with other exposures late in life, could increase the risk of PD and other neurological diseases (Costello et al., 2009). In addition, exposure to heavy metals is known to exacerbate neurological deficits and learning disabilities in children (Kozma, 2005) and is suspected of being associated with both onset and exacerbation of AD (Kotermanski and Johnson, 2009) and PD.

Note that the United States has seen an increasing trend in the occurrence of neurological diseases and deficits (Steenland et al., 2015). The factors involved with and affected by climate with particular implications for neurological functioning include malnutrition (Kar et al., 2008); exposure to hazardous chemicals, biotoxins, and metals in air, food, and water (Kozma, 2005); and changes in pest management (Handel, 2015). It goes without saying that understanding the role of climate in the incidence and progression of neurological conditions and how to prevent them is a critical need for public health and health care in the United States (Spellman, 2014).

SIDEBAR 12.5   ALGAL BLOOMS

Harmful algal blooms, or HAB, are sudden spurts of algal growth which can affect water quality adversely and indicate potentially hazardous changes in local water chemistry. Climate change is thought to play an increasingly important role in the spread of harmful algal blooms worldwide. Many HAB-related biotoxins cause significant neurotoxic effects in both animals and humans, including permanent neurological impairment. The algal diatom genus *Pseudo-nitzschia* produces domoic acid (DA), a potent naturally occurring marine neurotoxin that threatens the health of marine mammals, birds, and humans—acute severe DA is well defined as amnesic shellfish poisoning in people. Blooms of this algae have been increasing off the California coast, resulting in significant illness and death in marine animals. A decade of monitoring of health of California sea lions, a sentinel species for human health effects, indicates changes in neurologic symptomology and epidemiology of domoic acid toxicosis. In a study conducted by Goldstein et al. (2008), it was pointed out that three separate clinical syndromes are now present in exposed animals: acute domoic acid toxicosis with seizure, permanent hippocampal atrophy, and death; a second novel neurological syndrome characterized by epilepsy associated with the chronic consequences of sub-lethal exposure to domoic acid; and a third syndrome associated with *in utero* exposures resulting in permanent parturition, neonatal death, and significant neurotoxicity in the developing fetus developing in seizure activity as the animal grows, as well as long-lasting impacts on memory and learning. These observations indicate significant potential implications for human health effects, although their exact nature is not known and needs further study. Again, this is another area where we do not know what we do not know.

VECTORBORNE AND ZOONOTIC DISEASES

In the United States, diseases such as plague, typhus, malaria, yellow fever, and dengue fever, transmitted to humans by blood-feeding arthropods (i.e., mosquitoes, ticks, and fleas), were once common. The vectorborne pathogens include viruses, rickettsia, bacteria, protozoa, and worm parasites. Because of changes in land use, agricultural methods, and residential patterns, vector control and human behavior many of these diseases are on longer present. Note, however, we can't say the same about diseases that may be transmitted by birds or zoonoses (mammals) because the diseases continue to circulate in nature in many parts of the country. Most vectorborne diseases exhibit a distinct seasonal pattern, which clearly suggests that they

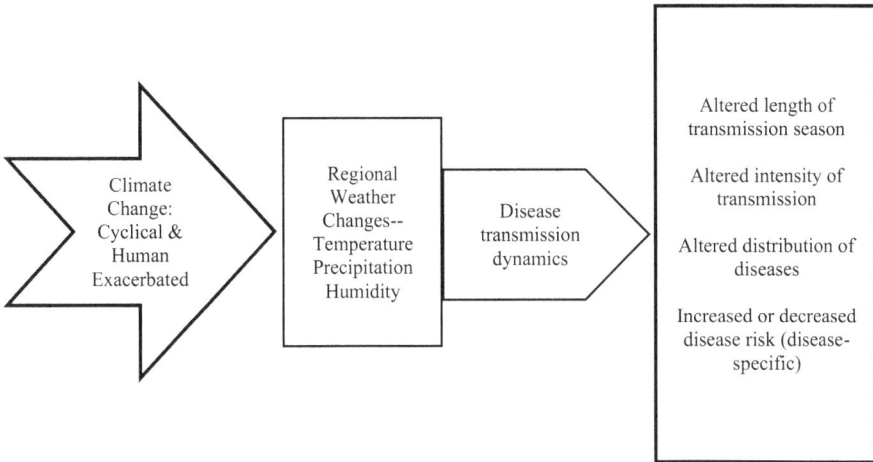

FIGURE 12.2   The transmission of disease may be affected by temperature, precipitation, and humidity.

Source: Adapted from Gubler et al. (2001).

are weather sensitive—climate sensitive—climate change sensitive. It is the weather variables—rainfall, temperature, and others—that affect in many ways both the vectors and the pathogens they transmit. A good example is the weather variable temperature, which can increase or reduce the survival rate, depending on the vector, its behavior, ecology, and many other factors.

SIDEBAR 12.6   CLIMATE CHANGE AND VECTORBORNE AND ZOONOTIC DISEASES

Note that climate is one of several factors that influence the distribution of vectorborne and zoonotic diseases (VBZDs) such as Lyme disease, Hantavirus, West Nile virus, and malaria. Climate change is a major concern with VBZD because certain environments will become more fitting for some VBZD, worsening their already substantial global burden, and possibly reintroducing some diseases into geographic areas where they had been previously wiped out. However, resurgence of vectorborne disease as a result of climate change is a major concern. This resurgence and reemergence may be exacerbated by severe increases in rainfall events and rainfall. Climate change–induced increases in temperature not only increase the altitude at which malaria transmission is possible but also intensify transmission at lower altitudes. Another factor to consider about climate change and vectorborne and zoonotic diseases is the changes in biodiversity of terrestrial and marine ecosystems, which in turn will alter the dynamics of predator–prey relationships, as well as vector and reservoir pathogen populations.

WATERBORNE DISEASES

Although the epidemiological relation between water and disease had been suggested as early as the 1850s, it was not until the establishment of the germ theory of disease by Pasteur in the mid-1880s that water as a carrier of disease-producing organisms was understood. And, in the 1850s, while London experienced the "Broad Street Well" cholera epidemic, Dr. John Snow conducted his now-famous epidemiological study. Dr. Snow concluded that the well had become contaminated by a visitor with the disease, who had arrived in the vicinity. Cholera was one of the first diseases to be recognized as capable of being waterborne. Also, this incident was probably the first reported disease epidemic attributed to direct recycling of non-disinfected water. Now, over 100 years later, the list of potential waterborne disease due to pathogens is considerably larger and includes bacterial, viral, and parasitic microorganisms, as shown in Tables 12.1, 12.2, and 12.3, respectively.

TABLE 12.1
Waterborne Diseases from Bacteria

Causative Agent	Disease
Salmonella typhosa	Typhoid fever
S. paratyphi	Paratyphoid fever
S. schottinulleri	
S. hirschfeldi C.	
Shigella flexneri	Bacillary dysentery
Sh. Dysenteriae	
Sh. Sonnel	
Sh. Paradysinteriae	
Vibrio comma V. cholerae	Cholera
Pasteurella tularensis	Tularemia
Brucella melitensis	Brucellosis
Leptospira icterchaemorrihagiae	Leptospirosis
Enteropathogenic E. coli	Gastroenteritis

TABLE 12.2
Waterborne Diseases from Human Enteric Viruses

Group	Subgroup
Eternovirus	Poliovirus
	Echovirus
	Coxsackie-virus
	A
	B
Reovirus	
Adenovirus	
Hepatitis	

TABLE 12.3
Waterborne Diseases from Parasites

Causative Agent	Symptoms
Ascario lumbricoides (round worm)	Ascariasis
Cryptosporidium muris and *parvum*	Cryptosporidiosis
Entamoeba histolytica	Amebiasis
Giardia lamblia	Giardiasis
Naegleria gruberi	Amoebic meningoecephalitis
Schistosoma mansoni	Schistosomiasis
Taenis saginata (beef tapeworm)	Taeniasis

A major cause for the number of disease outbreaks in potable water is contamination of the distribution system from cross-connections and back siphonage with non-potable water. However, outbreaks resulting from distribution system contamination are usually quickly contained and result in relatively few illnesses compared to contamination of the source water or a breakdown in the treatment system, which typically produce many cases of illnesses per incident. When considering the number of cases, the major causes of disease outbreaks are source water contamination and treatment deficiencies (White, 1992). Historically, about 46 percent of the outbreaks in the public water systems are found to be related to deficiencies in source water and treatment systems, with 92 percent of the causes of illness due to these two particular problems.

All natural waters support biological communities. Because some microorganisms can be responsible for public health problems, biological characteristics of the source water are one of the most important parameters in water treatment. In addition to public health problems, microbiology can also affect the physical and chemical water quality and treatment plant operation.

Pathogens of Primary Concern

Table 12.4 shows the attributes of three groups of pathogens of concern in water treatment, namely bacteria, viruses, and protozoa.

1. **Bacteria**—Bacteria are single-celled organisms typically ranging in size from 0.1 to 10 µm. Shape, components, size, and the manner in which they grow can characterize the physical structure of the bacterial cell. Most bacteria can be grouped by shape into four general categories: spheroid, rod, curved rod or spiral, and filamentous. Cocci, or spherical bacteria, are approximately 1 to 3 µm in diameter. Bacilli (rod-shaped bacteria) are variable in size and range from 0.3 to 1.5 µm in width (or diameter) and from 1.0 to 10.0 µm in length. Vibrios, or curved rod-shaped bacteria, typically vary in size from 0.6 to 1.0 µm in width (or diameter) and from 2 to 6 µm in length. Spirilla (spiral bacteria) can be found in lengths up to 50 µm, whereas filamentous bacteria can occur in length in excess of 100 µm.

2. **Viruses**—Viruses are microorganisms composed of the genetic material deoxyribonucleic acid (DNA) or ribonucleic acid (RNA) and a protective protein coat (either single, double, or partially double stranded). All viruses are obligate parasites, unable to carry out any form of metabolism and completely dependent upon host cells for replication. Viruses are typically 0.01 to 0.1 µm in size and are very species specific with respect to infection, typically attacking only one type of host. Although the principal modes of transmission for the hepatitis B virus and poliovirus are through food, personal contact, or exchange of body fluids, these viruses can be transmitted through potable water. Some viruses, such as the retroviruses (including HIV group), appear to be too fragile for water transmission to be a significant danger to public health (Spellman, 2007).

3. **Protozoa**—the Protozoa are single-cell eukaryotic microorganisms without cell walls that utilize bacteria and other organisms for food. Most protozoa are free living in nature and can be encountered in water; however, several species are parasitic and live on or in host organisms. Host organisms can vary from primitive organisms such as algae to highly complex organisms such as human beings. Several species of protozoa known to utilize human beings as hosts are shown in Table 12.5.

Recent Waterborne Disease Outbreaks

Within the past 40 years, several pathogenic agents never before associated with documented waterborne outbreaks have appeared in the United States. Enteropathogenic *E. coli* and *Giardia lamblia* were first identified to be the etiological agents responsible for waterborne outbreaks in the 1960s. The first recorded *Cryptosporidium*

TABLE 12.4
Attributes of the Three Waterborne Pathogens in Water Treatment

Organism	Size (µm)	Mobility	Point(s) of Origin	Resistance to Disinfection
Bacteria	0.1–10	Motile, nonmotile	Humans and animals, water, food	Type-specific bacterial spores typically have the highest resistance, whereas vegetative bacteria have the lowest resistance
Viruses	0.01–0.1	Nonmotile	Humans and animals, polluted water, contaminated food	Generally more resistant than vegetative bacteria
Protozoa	20-Jan	Motile, nonmotile	Humans and animals, sewage decaying vegetation, and water	More resistant than viruses or vegetative bacteria

TABLE 12.5
Human Parasitic Protozoans

Protozoan	Host(s)	Disease	Transmission
Acanathamoeba castellannii	Fresh water, sewage, humans, soil	Amoebic meningoencephalitis	Gains entry through abrasions, ulcers, and as secondary invader during other infections
Balantidium coli	Pigs, humans	Balantidiasis (dysentery)	Contaminated water
Cryptosporidium parvum	Animals, humans	Cryptosporidiosis	Person-to-person or animal-to-person contact, ingestion of fecally contaminated water or food, or contact with fecally contaminated environmental surfaces
Entamoeba histolytica	Humans	Amoebic dysentery	Contaminated water
Giardia lamblia	Animals, humans	Giardiasis (gastroenteritis)	Contaminated water
Naegleria fowleri	Soil, water, humans, and decaying vegetation	Primary amoebic meningoencephalitis	Nasal inhalation with subsequent penetration of nasopharynx; exposure from swimming in fresh-water lakes

infection in humans occurred in the mid-1970s. Also during that time was the first recorded outbreak of pneumonia caused by *Legionella pneumophila*. Recently, there have been numerous documented waterborne disease outbreaks that have been caused by *E. coli*, *G. lamblia*, *Cryptosporidium*, and *L. pneumophila*.

Escherichia coli

The first documented case of waterborne disease outbreak in the United States associated with enteropathogenic *E. coli* occurred in the 1960s. Various serotypes of *E. coli* have been implicated as the etiological agent responsible for disease in newborn infants, usually the result of cross-contamination in nurseries. Now, there have been several well-documented outbreaks of *E. coli* associated with adult waterborne disease. In 1975, the etiologic agent of a large outbreak at Crater Lake National Park was *E. coli* serotype 06:H16 (Craun, 1981).

Giardia lamblia

Similar to *Escherichia coli*, *Giardia lamblia* was first identified in the 1960s to be associated with waterborne outbreaks in the United States. Recall that *G. lamblia* is a flagellated protozoan that is responsible for Giardiasis, a disease that can range from being mildly to extremely debilitating. *Giardia* is currently one of the most commonly identified pathogens responsible for waterborne disease outbreaks. The life cycle of *Giardia* includes a cyst stage when the organism remains dormant and is extremely resilient (i.e., the cyst can survive some extreme environmental conditions). Once ingested by a warm-blooded animal, the life cycle of *Giardia* continues with excystation.

The cysts are relatively large (8–14 μm) and can be removed effectively by filtration using diatomaceous earth, granular media, or membranes. Giardiasis can be acquired by ingesting viable cysts from food or water or by direct contact with fecal material. In addition to humans, wild and domestic animals have been implicated as hosts. Between 1972 and 1981, fifty waterborne outbreaks of Giardiasis occurred, with about 20,000 reported cases (Craun and Jakubowski, 1996). Currently, no simple and reliable method exists to assay *Giardia* cysts in water samples. Microscopic methods for detection and enumeration are tedious and require examiner skill and patience. *Giardia* cysts are relatively resistant to chlorine, especially at higher pH and low temperatures.

Cryptosporidium

Cryptosporidium is a protozoan similar to *Giardia*. It forms resilient oocysts as part of its life cycle. The oocysts are smaller than *Giardia* cysts, typically about 4–6 μm in diameter. These oocysts can survive under adverse conditions until ingested by a warm-blooded animal and then continue with excystation. Due to the increase in the number of outbreaks of Cryptosporidiosis, a tremendous amount of research has focused on *Cryptosporidium* within the last 10 years. Medical interest has increased because of its occurrence as a life-threatening infection to individuals with depressed immune systems. As previously mentioned, in 1993, the largest documented waterborne disease outbreak in the United States occurred in Milwaukee and was determined to be caused by *Cryptosporidium*. An estimated 403,000 people became ill, 4,400 people were hospitalized, and 100 people died. The outbreak was associated with a deterioration in raw water quality and a simultaneous decrease in effectiveness of the coagulation-filtration process, which led to an increase in the turbidity of treated water and inadequate removal of *Cryptosporidium* oocysts.

Legionella pneumophila

An outbreak of pneumonia occurred in 1976 at the annual convention of the Pennsylvania American Legion. A total of 221 people were affected by the outbreak, and 35 of those afflicted died. The cause of pneumonia was not determined immediately despite an intense investigation by the Centers for Disease Control. Six months after the incident, microbiologists were able to isolate a bacterium from the autopsy lung tissue of one of the Legionnaires. The bacterium responsible for the outbreak was found to be distinct from other known bacterium and was named *Legionella pneumophila* (Witherell et al., 1988). Following the discovery of this organism, other *Legionella*-like organisms were discovered. Legionnaires' disease does not appear to be transferred person to person. Epidemiological studies have shown that the disease enters the body through the respiratory system. *Legionella* can be inhaled in water particles less than 5 μm in size from facilities such as cooling towers, hospital hot water systems, and recreational whirlpools.

Mechanism of Pathogen Inactivation

The three primary mechanisms of pathogen inactivation are:

- Destroy or impair cellular structural organization by attacking major cell constituents, such as destroying the cell wall or impairing the functions of semi-permeable membranes;
- Interfere with energy-yielding metabolism through enzyme substrates in combination with prosthetic groups of enzymes, thus rendering them non-functional; and
- Interfere with biosynthesis and growth by preventing synthesis of normal proteins, nucleic acids, coenzymes, or the cell wall.

Depending on the disinfectant and microorganism type, combinations of these mechanisms can also be responsible for pathogen inactivation. In water treatment, it is believed that the primary factors controlling disinfection efficiency are: (1) the ability of the disinfectant to oxidize or rupture the cell wall and (2) the ability of the disinfectant to diffuse into the cell and interfere with cellular activity (Montgomery, 1985). In addition, it is important to point out that disinfection is effective in reducing waterborne diseases because most pathogenic organisms are more sensitive to disinfection than are nonpathogens. However, disinfection is only as effective as the care used in controlling the process and ensuring that all of the water supply is continually treated with the amount of disinfectant required to produce safe water.

THE CLIMATE CHANGE ASSOCIATION

Note that there is a clear association between increases in precipitation and outbreaks of waterborne disease, both domestically and globally. Climate change is expected to produce more frequent and extreme precipitation worldwide. Curriero et al. (2001) point out that rainfall and runoff have been implicated in site-specific waterborne disease outbreaks, and because upward trends in heavy precipitation in the United States are projected to increase with climate change, we need to be aware of the potential consequences. Some of the largest outbreaks of waterborne disease in North America, particularly in the Great Lakes, have resulted after extreme rainfall events. For instance, in May 2000, heavy rainfall in Walkerton, Ontario, resulted in approximately 2,300 illnesses and 7 deaths after the town's drinking water became contaminated with *E. coli* 0157:H7 and *Campylobacter jejuni* (Hrudey et al., 2003). Note that there are 734 combined sewage and wastewater systems in and around the Great Lakes, with an estimated discharge of 850 billion gallons of untreated overflow water (USEPA, 2004). Employing a suite of seven climate change models to project extreme precipitation events in the Great Lakes region, scientists have been able to estimate the potential impact of climate change on waterborne disease rates (Patz et al., 2008).

SIDEBAR 12.7   THE "BUG" THAT MADE MILWAUKEE FAMOUS

It was in 1993, when the "bug—the pernicious parasite *Cryptosporidium*—made [itself and] Milwaukee famous (Mayo Foundation, 1996)." The *Cryptosporidium* outbreak in Milwaukee caused the deaths of 100 people—the largest episode of waterborne disease in the U.S. in the 70 years since health officials began tracking such outbreaks.

The massive waterborne outbreak in Milwaukee (more than 400,000 persons developed acute and often prolonged diarrhea or other gastrointestinal symptoms) increased interest in *Cryptosporidium* at an exponential level. The Milwaukee Incident spurred both public interest and the interest of public health agencies, agricultural agencies and groups, environmental agencies and groups, and suppliers of drinking water. This increased interest level and concern has spurred on new studies of *Cryptosporidium* with emphasis on developing methods for recovery, detection, prevention, and treatment (Fayer et al., 1997).

The USEPA has become particularly interested in this "new" pathogen. For example, in the reexamination of regulations on water treatment and disinfection, the USEPA issued MCLG and CCL for *Cryptosporidium*. The similarity to *Giardia lamblia* and the necessity to provide an efficient conventional water treatment capable of eliminating viruses at the same time forced the USEPA to regulate the surface water supplies in particular. The proposed "Enhanced Surface Water Treatment Rule" (ESWTR) included regulations from watershed protection to specialized operation of treatment plants (certification of operators and state overview) and effective chlorination. Protection against *Cryptosporidium* included control of waterborne pathogens such as *Giardia* and viruses (DeZuane, 1997).

Weather-Related Morbidity and Mortality

In the United States, there is nothing new about extreme weather events. This is especially the case in the past few decades. These extreme weather events, consisting of heat, cold, storms, floods, and lightning, have long been associated with morbidity and mortality. From 1940 to 2005, hurricanes caused approximately 4,300 deaths and flooding 7,000 deaths, primarily from injuries and drowning (Ashley and Ashley, 2007). As a result of changing weather patterns (via climate change), the frequency and intensity of all types of extreme weather events are expected to increase (Parry et al., 2007). The health impacts of these extreme weather events can be severe and include both direct impacts such as death and mental health effects and indirect impacts such as population displacement (migration to those so-called greener and safer pastures) and waterborne disease outbreak such as the1993 Milwaukee cryptosporidium outbreak caused by flooding that sickened an estimated 400,000 people (see Sidebar 12.7)

The bottom line: Climate change will force humans to negotiate with their changing environment or move on toward those greener pastures. The problem is that climate change is a global happening; therefore, the question becomes: will there be any greener pasture left to move on to if climate change continues and becomes more extreme and all encompassing?

THE SECOND EDGE OF THE SWORD

Apart from other unwanted effects, inequalities in adverse health effects across the globe are now generally accepted as being caused by climate change and being one of the primary drivers of displacement and migration of both humans and wildlife. There is a twist to this happening, however.

A twist?

Yes.

A twist in what or where or whatever?

The twist is a paradox—known by some as the Hispanic paradox or the Latino paradox. The truth of the matter is that many of the Hispanics/Latinos who migrate from below the southern border of the United States are not disease prone or disease carrying like many mistakenly believe. Moreover, most of the Hispanic migrants have health outcomes or health status that are "paradoxically" comparable to, or in frequent cases better than, those of their U.S. non-Hispanic White counterparts, even though Hispanics tend to have lower financial resources and less education. Worse population health and higher death rates everywhere in the world are usually associated with low economic status (Franzini, 2001). This is not the case with Hispanics, that is, with regard to health and morbidity rates. The paradox is that the migrant Hispanics are in better health than many of the fellow American Hispanics that they tend to settle with. Moreover, the sad part is that the healthy migrant Hispanics who spend any lengthy time in the United States start to pick up the ills of society, so to speak, and at the same time, their health declines to the point where they want to go back home. If they survive the journey back, it is likely that their health will continue to decline and end in their deaths.

NOTES

1. Based on F. Spellman (2009). *The Science of Air*, 2nd ed. Boca Raton, FL: CRC Press.
2. Based on information from *A Human Health Perspective On Climate Change* (2010). Accessed 3/1/22 @ www.niehs.gov/climatereport (a public domain document).

REFERENCES

Almond, D., and Currie, J. (2014). Killing me softly: The fetal origins hypothesis. *Journal of Economic Perspectives* 25(3):153–172.

Ashley, F., and Ashley W.S. (2007). Changes in US hurricane landscape. Accessed 12/15/21 @ https://www.fox13now.com/news/Local News.

Ashley, S.T., and Ashley, W.S. (2008). Flood fatalities in the United States. *Journal Meteorology and Climatology* 47(3):805–818.

Barker, D.J.P (1992). Fetal growth and adult disease. *International Journal of Obstetrics & Gynaecology* 99(4):275–276.

Barrett, B., Charles, J.W., and Temte, J.L. (2014). Climate change, human health, and epidemiological transition. *Prev. Med.* 70:69–75.

Bassil, K.L., Cole, D., Moineddin, R., and Craig, A.M. (2009). Temporal and spatial variation of heat-related illness using 911 medical dispatch data. *Environmental Research* 109(5):600–606.

Bates, B.C., Kundzewicz, Z.W., Shaohong, W., and Paluthof, J. (2008). *Climate change and water.* Geneva, Switzerland: IPCC.

Bukowski, J.A. (2001). Review of epidemiological evidence relating toluene to reproductive outcomes. *Regulatory Toxicology and Pharmacology: RTP* 33(2), april:147–156.

Burke, K.E., and Wei, H. (2009). Synergistic damage by UVA radiation and pollution. *Toxicol. and Health.* 25(4–5):219–224.

CDC (Centers for Disease Control and Prevention). (2022). More than 8 million babies worldwide are born with a serious birth defect each year. Accessed 03/07/2022 @ www.cdc.gov/ncbddd/birthdefects/index/html.

The CNA Corporation. (2007). National security and climate change. Accessed 03/09/2022 @ securityandclimate.cna.og.cav-files.

Correa-Villasenor, A. (2003). The Metropolitan Atlanta congential defects program. Accessed 12/12/21 @ coorreal@cdc.gov.

Costello, A., Abbas, M., Allen, A.A., Ball, S., Bell, S., Bellamy, R., Friel, S., Grace, N., Johnson, A., Kett, M., Lee, M., Levy, C., Maslin, M., McCoy, D., McGuire, B., Montgomery, H., Napier, D., Pagel, C., Patel, J., Puppim de Oliveira, J.A., Redclift, N., Rees, H., Rogger, D., Scott, J., Stepheson, J., Twigg, J., Wolff, J., and Patterson, C. (2009). Managing the health effects of climate change: Lancet and University College London Institute for Global Health Commission. *Lancet* 16(373):1693–733.

Craun, G.F. (1981). Outbreaks associated with recreational water in Crater Lake. Accessed 12/12/21 @ https://pubmied.ncbi.nih.gov/16175741.

Craun, G.F., and Jakubowski, W. (1996). *Status of Waterborne Giardiasis Outbreaks and Monitoring Methods.* Atlanta, GA: American Water Resources Association, Water Related Health Issue Symposium.

Curriero, F.C., Patz, J.A., Rose, J.B., and Lele, S. (2001). The association between extreme precipitation and waterborne disease outbreaks in the united States, 1948–1994. *Am. J. Public Health* 91(8):1194–1199.

D'Amato, G., and Cecchi, L. (2008). Effects of climate change on environmental factors in in respiratory allergic diseases. *Clinical and Experimental Allergy* 38(8):1264–1274.

Davido, A., Patzak, A., Dart, T., Sasler, M.P., Meraud, P., Masmoudi, R., Sembach, N., and Cao, T.H. (2006). Risk factors for heat related death during the August 2003 heat wave in Paris, France, in patients evaluated at the emergency department of the Hospital Europeen Georges Pompidou. Accessed @ alaian.davicodegp.ap-hop-paris.fr.

Davis, M.L., and Cornwell, D.A. (1991). *Introduction to Environmental Engineering.* New York: McGraw-Hill, Inc.

De Zuane, J. (1997). *Handbook of Drinking Water Quality.* New York: John Wiley & Sons, Inc.

Dong, S., Hwang, H.M., Shi, X., Holloway, L., and Yu, H. (2000). UVA-induced DNA single-strand cleavage by 1-hydroxypyrene and formation of covalent adducts between DNA and 1-hydrooypyrene. *Chem, Res. Toxicol.* 13(7):585–593.

Ebi, K., et al. (2008). Effects of global change on human health. In Gamble, J. et al., editors. *Analyses of the Effects of Global Change on Human Health and Welfare and Human Systems. A report by the U.S. Climate Change Science Program and the Subcommittee on Global Change Research.* Washington, DC: United States Environmental Protection Agency.

EPA. (2013). DDT: A brief history and status. Accessed 03/20/2013 @ www.epa.gov/pesticides/factsheets/chemicals/ddt-brief-history-status-htm.

Eskenazi, B., Rosas, L.G., Marks, A.R., Bradman, A., Harley, K., Holland, N., Johnson, C., Fenster, L., and Barr, D.B. (2008). Pesticide toxicity and the developing brain. *Basic & Clinical Pharmacology & Toxicology* 102(2):228–236.

Fayer, R., Speer, C.A., and Dudley, J.P. (1997). *The General Biology Cryptosporidium in Cryptosporidium and Cryptosporidiosis*. Fayer, R., editor. Boca Raton, FL: CRC Press.

Fouillet, A., Rey, G., Laurent, F., Pavillon, G., Bellee, S., Guihenneue-Jouyaux, C.J., Jougla, E., and Hemon, D. (2006). Excess mortality related to the August 2003 heat wave in France. *Int. Arch Occup. Environ. Health* 80:16–24.

Franzini, L.R. (2001). Widening of socioeconomic inequalities in U.S. death rate. Accessed 12/12/21 @ https://journals.plos.org/flosme/article?id=19.1371.

Fritze, J.G., Blaski, G.A., Burke, S., and Wiseman, J. (2008). Hope, despair, and transformation: Climate change and the Promotion of mental health and wellbeing. *International Journal of Mental Health* 2:1–10.

Gamble, J.I., and Ebi, K.L. (2008). *Analyses of the Effects of Global Change on Human Health and Welfare and Human Systems*. Washington, DC: United States Environmental Protection Agency.

Goldstein, T., Mazet, J.A.K., Zabka, T.S., Langlois, G., Colegrove, K.M., Silver, M., Bargu, S., Van Dolah, F., Leighfield, T., Conrad, P.A., Barakos, J., Williams, D.C., Dennison, S., Haulena, M., and Guiland, F.M.D. (2008). Novel symptomatology and changing epidemiology of domoic acid toxicosis in California in sea lions (*Zalophus californicus*): An increasing risk to marine mammal health. *Proc. Env. Sci.* 7(275):267–276.

Gregory, P.J., Johnson, S.H., Newton, A.C., and Ingram, L.S.L. (2009). Integrating pests and pathogens into the climate change/food security debate. *Journal of Environmental Biology* 60(10):2827–2838.

Gubler, D.J., Reiter, P., Ebi, K.L., Yap, W., Nasci, R., and Patz, J.A. (2001). Climate variability and change in the United States: Potential impacts on vector- and rodent-borne diseases. *Environmental Health Perspectives* 109(2):223–233.

Gundy, P.M., et. al. (2008). Survival of coronaviruses in water and wastewater. *Food and Environmental Virology* 1(1).

Handel, S.K. (2015). Handle pest control. Accessed 12/12/21 @ https://Canpages.CA/Business/SK/handle/pest.

Houghton, J.T., Ding, Y., Griggs, D.J., Noguer, M., van der Linden, P. J., Dai, X., Maskell, K., and Johnson, C.A. (2001). *Climate change 2001: The scientific basis*. New York, NY: Cambridge University Press.

Hrudey, S.E., Payment, P., Huck, P.M., Gillham, R.W., and Hrudey, E.J. (2003). A fatal waterborne disease epidemic in Walkerton, Ontario outbreaks in the developed world. *Water Sci. Technol* 47(3):7–14.

The Interdepartmental Committee for Atmospheric Science. (1975). *Ozone*. Washington, DC: Library of Congress Congressional Research Service.

IPCC Third Assessment Report: Synthesis Report. (2007). Accessed 01/26/2022 @ www.ipcc.ch/report/ar5/syr/.

Kar, B.R., Rao, S.L., and Chandramouli, B.A. (2008). Cognitive development in children with chronic protein energy malnutrition. 4:31.

Kotermanski, S.E., and Johnson, J.W. (2009). Mg^{2+} imparts NMDA receptor subtype selectivity to the Alzheimer's drug memantine. *Journal of Neuroscience* 29(9):2774–2779.

Kovats, R.S., and Hajat, S. (2008). Heat stress and public health: A critical review. *Annual Review of Public Health 2008* 29:41–55.

Kozma, C. (2005). Neonatal toxicity and transient neurodevelopmental deficits following pre-natal exposure to lithium: Another critical report and a review of the literature. *Am J Med Genet A* 1;132A(4):441–444.

Lynch, S.M., Mahajan, R., Freeman, L.E.B., Hoppin, J.A., and Alavanja, M.C.R. (2009). Cancer incidence among pesticide applicators exposed to butylate in the Agricultural Health Study (AHS). *Environmental Research* 109(7):860–868.

Mayo Foundation. (1996). *The "Bug" That Made Milwaukee Famous*. Rochester, MN: Mayo Foundation.

McAloose, D., and Newton, A.L. (2009). Wildlife cancer: A conservation perspective. *Nat Rev Cancer* 9(7):517–526.

McLaughlin, J.B., DePaola, A., Bopp, C.A., Martinek, K.A., Napolilli, N.P., Allison, C.G., Murray, S.L., Thompson, E.C., Bird, M.M., and Middaugh, J.P. (2005). Outbreak of *Vibrio parahaemolyticus* gastroenteritis associated with Alaskan oysters. *N. Engl. J. Med.* 353:14.

Medema, G., et. al. (2020). Presence of SARS-cononavirus-2 RNA in sewage and corre-lation with reported COVID-19 prevalence in the early stage of the epidemic in the Netherlands. *Environmental Science & Technology Letters* 7(7):511–516.

Meehl, G.G., and Tebaldi, C. (2004). More intense, more frequent, and more longer lasting heat waves in the 21st century. *Science* 305(5686):994–997.

Montgomery, J.M. (1985). *Water Treatment and Principles*. New York: Wiley & Sons.

National Cancer Institute. (2009). What is cancer? Accessed 03/01/2022 @ www.cancer.gov/cancertopics/wahat-is-cancer.

Okoh, A.I., et. al. (2010). Inadequately treated wastewater as a source of human enteric viruses in the environment. *International Journal of Environmental Research and Public Health* 7(6):2620–2637.

Parry, M.L., Canziani, O.F., Paluntikof, J.P., van der Linden, P.J., and Hanson, C.E. (eds.) (2007). *Contribution of Working Group II to the Fourth Assessment Report of Intergovernmental Panel on Climate Change, 2007*. Cambridge, UK and New York, NY: Cambridge University Press.

Patz, J.A., Vavrus, S.J., Uejio, C.K., and McLellan, S.L. (2008). Climate change and water-borne disease risk in the Great Lakes region of the U.S. *Am J Prev Med.* 35(5):451–458.

Petronis, A. (2010). Epigenetics as a unifying principle in the aetiology of complex traits and disease. *Nature* 465:721–727.

Pradat, P., Robert-Gnansia, E., Di Tanna, G.L., Rosano, A., Rosano, A., Lisi, A., and Mastroiacovo, P. (2003). First trimester exposure to corticosteroids and oral clefts. Accessed 3/7/2022 @ https://pubmed.ncbi,nim.nih.gov/14745915.

Schimmelppfennig, D., Lewandrowski, J., Reilly, J., Tsigas, M., and Perry, I. (2021). Agricultural adaptation to climate change: Issues of long run sustainability. Accessed 03/03/2022 @ htpps://www.ers.usda.gov/webdocs/publications/40712/17912-aer740a.

Smolinski, M.S., Hamburg, M.A., and Lederberg, J. (eds.) (2003). *Sokurenko, E.V., Goum.* Washington, DC: National Academies Press.

Sokurenko, E.V., Gomulkiewicz, R., and Dykhuizen, D.E. (2006). Source-sink dynamics of virulence evolution. *Nat. Rev. Microbiol.* 4(7):548–555.

Spellman, F.R. (1998). *The Science of Air*. Boca Raton, FL: CRC Press.

Spellman, F.R. (1999). *The Science of Environmental Pollution*. Boca Raton, FL: CRC Press.

Spellman, F.R. (2007). *Ecology for Non-Ecologists*. Lanham, MD: Government Institutes Press.

Spellman, F.R. (2014). *Personal Care Products and Pharmaceuticals in Wastewater and the Environment*. Lancaster, PA: DESTEC Press.

Spellman, F.R., and Stoudt, M. (2013). *Environmental Health*. Lanham, MD: Bernan Press.

Steenland, K., Goldstein, F.C., Levey, A.I., and Wharton, W. (2015). A meta-analysis of Alzheimer's disease incidence and prevalence comparing African-Americans and Caucasians. *Journal of Alzheimer's Disease* 50(1).

Tucker, M.A. (2009). Melanoma epidemiology. *Hematol. Oncol. Clin North Am.* 23(3):383–395.

USEPA. (1998). *Research Plan for Endocrine Disruptors*. Washington, DC: United States Environmental Protection Agency.

USEPA. (2004). Characterization of CSOs and SSOs. Accessed 3/12/22 Climate change and human skin cancer. *Photochem. Photobio. Sci.* 7(6):730–733.

van der Leun, J.C., and Place, R. (2008). Climate change and human skin cancer. Accessed @ htttps://pubmed, ncbi.nlm.nih.gov.

Wadhwa, P.D., Buss, C., Entringer, S., and Swanson, J.M. (2009). Developmental origins of health and disease: Brief history of the approach and current focus on epigenetic mechanisms. *Semin. Reprod. Med.* 27(5):358–368.

White, G.C. (1992). *Handbook of Chlorination and Alternative Disinfectants*. New York: Van Nostrand Reinhold.

WHO. (2008). *Global Malaria Programme, World Malaria Report 2008*. Geneva: World Health Organization.

Witherell, L.E., et al. (1988). Investigation of *Legionella pneumophila* in drinking water. *J. AWWA*. 80(2):88–93.

Workman, E. (2021). Intensification of heatwaves in the Middle East and North Africa could be a significant driver of migration from the region. Accessed 03/06/2022 @ https://earthrefuge.org/intensification-of-heatwaves-in-the-middle-east-and-north-africa-could-be-significant-driver-of-migration-from-the-region.

Zurer, P.S. (1988). Studies on ozone destruction expand beyond Antarctic. *C & E News* May:18–25.

SUGGESTED READINGS

Correa-Villaseno, A., Cragan, J., Kurik, J., O'Leary, L., Siffel, C., and Williams, L. (2003). The Metropolitan Atlanta Congenital Defects Program: 35 years of birth defects surveillance at the centers for disease control and prevention. *Birth Defects Research Part A: Clinical and Molecular Teratology* 67(9):617–624.

Sleijffers, A., Garssen, J., Vos, J.G., and van Loveren, H. (2004). Ultraviolet light and resistance to infectious disease. *Journal of Immunotoxicology* 1(1):3–14.

Spellman, F.R., and Whiting, N. (2006). *Environmental Science and Technology: Concepts and Applications*. Boca Raton, FL: CRC Press.

Glossary

A

Abrupt climate change: sudden (on the order of decades) large changes in some major component of the climate system, with rapid, widespread effects.

Absorption: any process by which one substance penetrates the interior of another substance.

Acid: has a pH of water less than 5.5; pH modifier used in the U.S. Fish and Wildlife Service wetland classification system; in common usage, acidic water has a pH less than 7.

Acidic deposition: the transfer of acidic or acidifying substances from the atmosphere to the surface of the Earth or to objects on its surface. Transfer can be either by wet deposition processes (rain, snow, dew, fog, frost, hail) or by dry deposition (gases, aerosols, or fine to coarse particles).

Acid rain: precipitation with higher-than-normal acidity, caused primarily by sulfur and nitrogen dioxide air pollution.

Acre-foot (acre-ft.): the volume of water needed to cover an acre of land to a depth of 1 foot; equivalent to 43,560 cubic feet or 32,851 gallons.

Activated carbon: a very porous material that after being subjected to intense heat to drive off impurities can be used to adsorb pollutants from water.

Adaptation: process of adjustment to actual or expected climate change and its effects. Adaptation seeks to moderate or avoid harm or exploit beneficial opportunities. In some natural systems, human intervention may facilitate adjustment to expected climate change and its effects.

Adaptive capacity: ability of systems, institutions, humans, and other organisms to adjust to potential damage, take advantage of opportunities, and respond to consequences of climate impacts.

Adsorption: the process by which one substance is attracted to and adheres to the surface of another substance without actually penetrating its internal structure.

Aeration: a physical treatment method that promotes biological degradation of organic matter. The process may be passive (when waste is exposed to air), or active (when a mixing or bubbling device introduces the air).

Aerobic bacteria: a type of bacteria that requires free oxygen to carry out metabolic function.

Aerosols: small particles or liquid droplets in the atmosphere that can absorb or reflect sunlight depending on their composition.

Afforestation: planting of new forests on lands that historically have not been forests (IPCC Third Assessment Report Working Group I: The Scientific Basis).

Albedo: the amount of solar radiation reflected from an object or surface, often expressed as a percentage.

Algae: chlorophyll-bearing nonvascular, primarily aquatic species that have no true roots, stems, or leaves; most algae are microscopic, but some species can be as large as vascular plants.

Algal bloom: the rapid proliferation of passively floating, simple plant life, such as blue-green algae, in and on a body of water.

Alkaline: has a pH greater than 7; pH modifier in the U.S. Fish and Wildlife Service wetland classification system; in common usage, a pH of water greater than 7.4.

Alluvial aquifer: a water-bearing deposit of unconsolidated material (sand and gravel) left behind by a river or other flowing water.

Alluvium: general term for sediments of gravel, sand, silt clay, or other particulate rock material deposited by flowing water, usually in the beds of rivers and streams, on a flood plain, on a delta, or at the base of a mountain.

Alpine snow glade: a marshy clearing between slopes above the timberline in mountains.

Alternative energy: made by people or resulting from human activities. Usually used in the context of emissions that are produced as a result of human activities (NASA's Earth Observatory library).

Amalgamation: the dissolving or blending of a metal (commonly gold and silver) in mercury to separate it from its parent material.

Ammonia: a compound of nitrogen and hydrogen (NH_3) that is a common by-product of animal waste. Ammonia readily converts to nitrate in soils and streams.

Anaerobic: pertaining to, taking place in, or caused by the absence of oxygen.

Anomalies: as related to fish, externally visible skin, or subcutaneous disorders, including deformities, eroded fins, lesions, and tumors.

Anthropogenic: having to do with or caused by humans.

Anticline: a fold in the Earth's crust, convex upward, whose core contains stratigraphically older rocks.

Aquaculture: the science of faming organisms that live in water, such as fish, shellfish, and algae.

Aquatic: living or growing in or on water.

Aquatic guidelines: specific levels of water quality which, if reached, may adversely affect aquatic life. These are nonenforceable guidelines issued by a governmental agency or other institution.

Aquifer: a geologic formation, group of formations, or part of a formation that contains sufficient saturated permeable material to yield significant quantities of water to springs and wells.

Aquitard: a saturated, but poorly permeable, geologic unit that impedes groundwater movement and does not yield water freely to wells but may transmit appreciable water to and from adjacent aquifers and, where sufficiently thick, may constitute an important groundwater storage unit. Areally extensive aquitards may function regionally as confined units within aquifer systems.

Arroyo: a small, deep, flat-floored channel or gully of an ephemeral or intermittent stream, usually with nearly vertical banks cut into unconsolidated material.

Artesian: an adjective referring to confined aquifers. Sometimes the term artesian is used to denote a portion of a confined aquifer where the altitude of the potentiometric surface is above the land surface (flowing wells and artesian wells are synonymous in this usage). But more generally the term indicates that the altitudes of the potentiometric surface are above the altitude of the base or the confining unit (artesian wells and flowing wells are not synonymous in this case.

Artificial recharge: augmentation of natural replenishment of groundwater storage by some method of construction, spreading of water, or pumping water directly into an aquifer.

Atmosphere: the gaseous envelope surrounding the Earth. Its components, characteristics, and properties are detailed in the following.

Note: Much of the information pertaining to atmospheric gases that follows was adapted from the Compressed Gas Association's *Handbook of Compressed Gases* (1990) and *Environmental Science and Technology: Concepts and Applications* (2006).

ATMOSPHERIC NITROGEN

Nitrogen (N_2) makes up the major portion of the atmosphere (78.03 percent by volume, 75.5 percent by weight). It is a colorless, odorless, tasteless, nontoxic, and almost totally inert gas. Nitrogen is nonflammable, will not support combustion, and is not life supporting. "Not life supporting?" No, gaseous nitrogen is not. The obvious question becomes: If gaseous nitrogen does not support life, what is it doing in our atmosphere—what good is it? Logical question. However, the question is incorrect; it implies something that is not true: nitrogen is indeed good and more. Without nitrogen, we could not survive.

Nitrogen is part of earth's atmosphere primarily because, over time, it has simply accumulated in the atmosphere and remained in place and in balance. This nitrogen accumulation process has occurred because, chemically, nitrogen is not very reactive. When released by any process, it tends not to recombine with other elements and accumulates in the atmosphere. And this is a good thing because we need nitrogen. No, we don't need it for breathing, but we need it for other life-sustaining processes.

Let's take a look at a couple of reasons gaseous nitrogen is so important to us. Although nitrogen in its gaseous form is of little use to us, after oxygen, carbon, and hydrogen, it is the most common element in living tissues. As a chief constituent of chlorophyll, amino acids, and nucleic acids—the "building blocks" of proteins (which are used as structural components in cells)—nitrogen is essential to life. Nitrogen is dissolved in and is carried by the blood. Nitrogen does not appear to enter into any chemical combination as it is carried throughout the body. Each time we breathe, the same amount of nitrogen is exhaled as is inhaled. Animals cannot use nitrogen directly but only when it is obtained by eating plant or animal tissues; plants obtain the nitrogen they need when it is in the form of inorganic compounds, principally nitrate and ammonium.

Gaseous nitrogen is converted to a form usable by plants (nitrate ions) chiefly through the process of nitrogen fixation via the nitrogen cycle.

Via the *nitrogen cycle*, aerial nitrogen is converted into nitrates mainly by microorganisms, bacteria, and blue-green algae. Lightning also converts some aerial nitrogen gas into forms that return to the earth as nitrate ions in rainfall and other types of precipitation. Ammonia plays a major role in the nitrogen cycle. Excretion by animals and anaerobic decomposition of dead organic matter by bacteria produce ammonia.

Ammonia, in turn, is converted by nitrification bacteria into nitrites and then into nitrates. This process is known as nitrification. Nitrification bacteria are aerobic. Bacteria that convert ammonia into nitrites are known as nitrite bacteria (*Nitrosococcus* and *Nitrosomonas*). Although nitrite is toxic to many plants, it usually does not accumulate in the soil. Instead, other bacteria (such as *Nitrobacter*) oxidize the nitrite to form nitrate (NO_3^-), the most common biologically usable form of nitrogen.

Nitrogen reenters the atmosphere through the action of denitrifying bacteria, which are found in nutrient-rich habitats such as marshes and swamps. These bacteria break down nitrates into nitrogen gas and nitrous oxide (N_2O), which then reenter the atmosphere. Nitrogen also reenters the atmosphere from exposed nitrate deposits, emissions from electric power plants, automobiles, and volcanoes.

Nitrogen: Physical Properties

The physical properties of nitrogen are noted in Table G.1.

TABLE G.1
Nitrogen: Physical Properties

Chemical formula	N2
Molecular weight	28.01
Density of gas @ 70°F	0.072 lb/ft^3
Specific gravity of gas @ 70°F & 1 atm (air = 1)	0.967
Specific volume of gas @ 70°F & 1 atm	13.89 ft^3
Boiling point @ 1 atm	−320.4°F
Melting point @ 1 atm	−345.8°F
Critical temperature	−232.4°F
Critical pressure	493 psia
Critical density	19.60 lb/ft^3
Latent heat of vaporization @ boiling point	85.6 Btu/lb
Latent heat of fusion @ melting point	11.1 Btu/lb

Nitrogen: Uses

In addition to being the preeminent (with regard to volume) component of Earth's atmosphere and providing an essential ingredient in sustaining life, nitrogen gas has many commercial and technical applications. As a gas, it is used in heat-treating primary metals; the production of semi-conductor electronic components as a blanketing atmosphere; blanketing of oxygen-sensitive liquids and volatile liquid chemicals; inhibition of aerobic bacteria growth; and the propulsion of liquids through canisters, cylinders, and pipelines.

Nitrogen Oxides

There are six oxides of nitrogen: nitrous oxide (N_2O), nitric oxide (NO), dinitrogen trioxide (N_2O_3), nitrogen dioxide (NO_2), dinitrogen tetroxide (N_2O_4), and dinitrogen pentoxide (N_2O_5).

Nitric oxide, nitrogen dioxide, and nitrogen tetroxide are fire gases. One or more of them is generated when certain nitrogenous organic compounds (polyurethane) burn. Nitric oxide is the product of incomplete combustion, whereas a mixture of nitrogen dioxide and nitrogen tetroxide is the product of complete combustion.

The nitrogen oxides are usually collectively symbolized by the formula NO_x. USEPA, under the Clean Air Act (CAA), regulates the amount of nitrogen oxides that commercial and industrial facilities may emit to the atmosphere. The primary and secondary standards are the same: The annual concentration of nitrogen dioxide may not exceed 100 µg/m³ (0.05 ppm).

ATMOSPHERIC OXYGEN

Oxygen (O_2—Greek *oxys* "acid"; *genes* "forming") constitutes approximately a fifth (21 percent by volume and 23.2 percent by weight) of the air in Earth's atmosphere. Gaseous oxygen (O_2) is vital to life as we know it. On Earth, oxygen is the most abundant element. Most oxygen on Earth is not found in the free state but in combination with other elements as chemical compounds. Water and carbon dioxide are common examples of compounds that contain oxygen, but there are countless others.

At ordinary temperatures, oxygen is a colorless, odorless, tasteless gas that not only supports life but also combustion. All the elements except the inert gases combine directly with oxygen to form oxides. However, oxidation of different elements occurs over a wide range of temperatures.

Oxygen is nonflammable, but it readily supports combustion. All materials that are flammable in air burn much more vigorously in oxygen. Some combustibles, such as oil and grease, burn with nearly explosive violence in oxygen if ignited.

Oxygen: Physical Properties

The physical properties of oxygen are noted in Table G.2

TABLE G.2
Oxygen: Physical Properties

Chemical formula	O_2
Molecular weight	31.9988
Freezing point	−361.12°F
Boiling point	−297.33°F
Heat of fusion	5.95 Btu/lb
Heat of vaporization	91.70 Btu/lb
Density of gas @ boiling point	0.268 lb/ft³
Density of gas @ room temperature	0.081 lb/ft³
Vapor density (air = 1)	1.105
Liquid-to-gas expansion ratio	875

Oxygen: Uses

The major uses of oxygen stem from its life-sustaining and combustion-supporting properties. It also has many industrial applications (when used with other fuel gases such as acetylene) including metal cutting, welding, hardening, and scarfing.

Ozone: Just Another Form of Oxygen

Ozone (O_3) is a highly reactive pale-blue gas with a penetrating odor. Ozone is an allotropic modification of oxygen. An allotrope is a variation of an element that possesses a set of physical and chemical properties significantly different from the "normal" form of the element. Only a few elements have allotropic forms; oxygen, phosphorous, and sulfur are some of them. Ozone is just another form of oxygen. It is formed when the molecule of the stable form of oxygen (O_2) is split by ultraviolet radiation or electrical discharge; it has three instead of two atoms of oxygen per molecule. Thus, its chemical formula is represented by O_3.

Ozone forms a thin layer in the upper atmosphere, which protects life on Earth from ultraviolet rays, a cause of skin cancer. At lower atmospheric levels, it is an air pollutant and contributes to the greenhouse effect. At ground level, ozone, when inhaled, can cause asthma attacks, stunted growth in plants, and corrosion of certain materials. It is produced by the action of sunlight on air pollutants, including car exhaust fumes, and is a major air pollutant in hot summers.

ATMOSPHERIC CARBON DIOXIDE

Carbon dioxide (CO_2) is a colorless, odorless gas (although it is felt by some persons to have a slight pungent odor and biting taste), slightly soluble in water and denser than air (one and half times heavier than air), and slightly acid. Carbon dioxide gas is relatively non-reactive and nontoxic. It will not burn, and it will not support combustion or life.

CO_2 is normally present in atmospheric air at about 0.035 percent by volume and cycles through the biosphere (carbon cycle). Carbon dioxide, along with water vapor, is primarily responsible for the absorption of infrared energy re-emitted by the Earth and, in turn, some of this energy is reradiated back to the Earth's surface. It is also a normal end product of human and animal metabolism. The exhaled breath contains up to 5.6 percent carbon dioxide. In addition, the burning of carbon-laden fossil fuels releases carbon dioxide into the atmosphere. Much of this carbon dioxide is absorbed by ocean water, some of it is taken up by vegetation through photosynthesis in the carbon cycle, and some remains in the atmosphere. Today, it is estimated that the concentration of carbon dioxide in the atmosphere is approximately 350 parts per million (ppm) and is rising at a rate of approximately 20 ppm every decade. The increasing rate of combustion of coal and oil has been primarily responsible for this occurrence, which (as we will see later in this text) may eventually have an impact on global climate.

Carbon Dioxide: Physical Properties

The physical properties of carbon dioxide are noted in Table G.3.

TABLE G.3
Carbon Dioxide: Physical Properties

Chemical formula	CO_2
Molecular weight	44.01
Vapor pressure @ 70°F	838 psig

Density of the gas @ 70°F & 1 atm	0.1144 lb/ft³
Specific gravity of the gas @ 70°F & 1 atm (air = 1)	1.522
Specific volume of the gas @ 70°F & 1 atm	8.741 ft³/lb
Critical temperature	−109.3°F
Critical pressure	1,070.6 psia
Critical density	29.2 lb/ft³
Latent heat of vaporization @ 32°F	100.8 Btu/lb
Latent heat of fusion @ −69.9°F	85.6 Btu/lb

Carbon Dioxide: Uses

Solid carbon dioxide is used quite extensively to refrigerate perishable foods while in transit. It is also used as a cooling agent in many industrial processes, such as grinding, rubber work, cold-treating metals, vacuum cold traps, and so on.

Gaseous carbon dioxide is used to carbonate soft drinks, for pH control in water treatment, in chemical processing, as a food preservative, and in pneumatic devices.

Atmospheric Argon

Argon (Ar—Greek *argos* "idle") is a colorless, odorless, tasteless, nontoxic, nonflammable gaseous element (noble gas). It constitutes almost 1 percent of the Earth's atmosphere and is plentiful compared to the other rare atmospheric gases. It is extremely inert, and forms no known chemical compounds. It is slightly soluble in water.

Argon: Physical Properties

The physical properties of argon are noted in Table G.4.

TABLE G.4
Argon: Physical Properties

Chemical formula	Ar
Molecular weight	39.95
Density of the gas @ 70°F	0.103 lb/ft³
Specific gravity of the gas @ 70°F	1.38
Specific volume of the gas @ 70°F	9.71 ft³/lb
Boiling point at 1 atm	−302.6°F
Melting point at 1 atm	−308.6°F
Critical temperature	−188.1°F
Critical pressure	711.5 psia
Critical density	33.444 lb/ft³
Latent heat of vaporization @ boiling point and 1 atm	69.8 But/lb
Latent heat of fusion	12.8 Btu/lb

Argon: Uses

Argon is used extensively in filling incandescent and fluorescent lamps and electronic tubes; to provide a protective shield for growing silicon and germanium crystals; and as a blanket in the production of titanium, zirconium, and other reactive metals.

ATMOSPHERIC NEON

Neon (Ne—Greek *neon* "new") is a colorless, odorless, gaseous, nontoxic, chemically inert element. Air is about 2 parts per thousand neon by volume.

Neon: Physical Properties

The physical properties of neon are noted in Table G.5.

TABLE G.5
Neon: Physical Properties

Chemical formula	Ne
Molecular weight	20.183
Density of the gas @ 70°F & 1 atm	0.05215 lb/ft³
Specific gravity of the gas @ 70°F & 1 atm	0.696
Specific volume of the gas @ 70°F & 1 atm	19.18 ft³/lb
Boiling point at 1 atm	−410.9°F
Melting point at 1 atm	−415.6°F
Critical temperature	−379.8°F
Critical pressure	384.9 psia
Critical density	30.15 lb/ft³
Latent heat of vaporization @ boiling point	37.08 But/lb
Latent heat of fusion	7.14 Btu/lb

Neon: Uses

Neon is used principally to fill lamp bulbs and tubes. The electronics industry uses neon singly or in mixtures with other gases in many types of gas-filled electron tubes.

ATMOSPHERIC HELIUM

Helium (He—Greek *helios* "Sun") is inert (and as a result does not appear to have any major effect on, or role in, the atmosphere), nontoxic, odorless, tasteless, non-reactive, forms no compounds, colorless and makes up about 0.00005 percent (5 ppm) by volume of air in the Earth's atmosphere. Helium, as with neon, krypton, hydrogen, and xenon, is a noble gas. Helium is the second-lightest element; only hydrogen is lighter. It is one seventh as heavy as air. Helium is nonflammable and is only slightly soluble in water.

Helium: Physical Properties
The physical properties of helium are noted in Table G.6.

TABLE G.6
Helium: Physical Properties

Chemical formula	He
Molecular weight	4.00
Density of the gas @ 70°F & 1 atm	0.0103 lb/ft^3
Specific gravity of the gas @ 70°F & 1 atm	0.138
Specific volume of the gas @ 70°F & 1 atm	97.09 ft^3/lb
Boiling point @ 1 atm	−452.1°F
Critical temperature	−450.3°F
Critical pressure	33.0 psia
Critical density	4.347 lb/ft^3
Latent heat of vaporization @ boiling point & 1 atm	8.72 But/lb

---

ATMOSPHERIC KRYPTON

Krypton (Kr—Greek *kryptos* "hidden") is a colorless, odorless, inert gaseous component of Earth's atmosphere. It is present in very small quantities in the air (about 114 parts per million—ppm).

Krypton: Physical Properties
The physical properties of krypton are noted in Table G.7.

TABLE G.7
Krypton: Physical Properties

Chemical formula	Kr
Molecular weight	83.80
Density of the gas @ 70°F & 1 atm	0.2172 lb/ft^3
Specific gravity of the gas @ 70°F & 1 atm	2.899
Specific volume of the gas @ 70°F & 1 atm	4.604 ft^3/lb
Boiling point @ 1 atm	−244.0°F
Melting point @ 1 atm	−251°F
Critical temperature	−82.8°F
Critical pressure	798.0 psia
Critical density	56.7 lb/ft^3
Latent heat of vaporization @ boiling point	46.2 Btu/lb
Latent heat of fusion	8.41 But/lb

Krypton: Uses

Krypton is used principally to fill lamp bulbs and tubes. The electronics industry uses it singly or in mixtures in many types of gas-filled electron tubes.

ATMOSPHERIC XENON

Xenon (Xe—Greek *xenon* "stranger") is a colorless, odorless, nontoxic, inert, heavy gas that is present in very small quantities in the air (about 1 part in 20 million).

Xenon: Physical Properties

The physical properties of xenon are noted in Table G.8.

TABLE G.8
Xenon: Physical Properties

Chemical formula	Xe
Molecular weight	131.3
Density of the gas @ 70°F & 1 atm	0.3416 lb/ft³
Specific gravity of the gas @ 70°F & 1 atm	4.560
Specific volume of the gas @ 70°F & 1 atm	2.927 ft³/lb
Boiling point at 1 atm	−162.6°F
Melting point at 1 atm	−168°F
Critical temperature	61.9°F
Critical pressure	847.0 psia
Critical density	68.67 lb/ft³
Latent heat of vaporization at boiling point	41.4 Btu/lb
Latent heat of fusion	7.57 Btu/lb

Xenon: Uses

Xenon is used principally to fill lamp bulbs and tubes. The electronics industry uses it singly or in mixtures in many types of gas-filled electron tubes.

ATMOSPHERIC HYDROGEN

Hydrogen (H_2—Greek *hydros* + *gen* "water generator") is a colorless, odorless, tasteless, nontoxic, flammable gas. It is the lightest of all the elements and occurs on Earth chiefly in combination with oxygen as water. Hydrogen is the most abundant element in the universe, where it accounts for 93 percent of the total number of atoms and 76 percent of the total mass. It is the lightest gas known, with a density approximately 0.07 that of air. Hydrogen is present in the atmosphere, occurring in concentrations of only about 0.5 ppm by volume at lower altitudes.

Hydrogen: Physical Properties

The physical properties of hydrogen are noted in Table G.9.

TABLE G.9
Hydrogen: Physical Properties

Chemical formula	H_2
Molecular weight	2.016
Density of the gas @ 70°F & 1 atm	0.00521 lb/ft^3
Specific gravity of the gas @ 70° & 1 atm	0.06960
Specific volume of the gas @ 70°F & 1 atm	192.0 ft^3/lb
Boiling point @ 1 atm	−423.0°F
Melting point @ 1 atm	−434.55°F
Critical temperature	−399.93°F
Critical pressure	190.8 psia
Critical density	1.88 lb/ft^3
Latent heat of vaporization @ boiling point	191.7 Btu/lb
Latent heat of fusion	24.97 Btu/lb

Hydrogen: Uses

Hydrogen is used by refineries, petrochemical, and bulk chemical facilities for hydro-treating, catalytic reforming, and hydro-cracking. Hydrogen is used in the production of a wide variety of chemicals. Metallurgical companies use hydrogen in the production of their products. Glass manufacturers use hydrogen as a protective atmosphere in a process whereby molten glass is floated on a surface of molten tin. Food companies hydrogenate fats, oils, and fatty acids to control various physical and chemical properties. Electronic manufacturers use hydrogen at several steps in the complex processes for manufacturing semiconductors.

ATMOSPHERIC WATER

Leonardo da Vinci understood the importance of water when he said: "Water is the driver of nature." da Vinci was actually acknowledging what most scientists and many of the rest of us have come to realize: Water, propelled by the varying temperatures and pressures in Earth's atmosphere, allows life as we know it to exist on our planet (Gradel & Crutzen, 1995).

The water vapor content of the lower atmosphere (troposphere) is normally within a range of 1–3 percent by volume, with a global average of about 1 percent. However, the percentage of water in the atmosphere can vary from as little as 0.1 percent or as much as 5 percent water, depending upon altitude; water in the atmosphere decreases with increasing altitude. Water circulates in the atmosphere in the hydrologic cycle.

Water vapor contained in Earth's atmosphere plays several important roles: (1) it absorbs infrared radiation; (2) it acts as a blanket at night, retaining heat from the Earth's surface; and (3) it affects the formation of clouds in the atmosphere.

ATMOSPHERIC PARTICULATE MATTER

There are significant numbers of particles (particulate matter) suspended in the atmosphere, particularly the troposphere. These particles originate in nature from smokes, sea sprays, dusts, and the evaporation of organic materials from vegetation. There is also a wide variety of nature's living or semi-living particles—spores and pollen grains, mites and other tiny insects, spider webs, and diatoms. The atmosphere also contains a bewildering variety of anthropogenic (human-made) particles produced by automobiles, refineries, production mills, and many other human activities.

Atmospheric particulate matter varies greatly in size (colloidal-sized particles in the atmosphere are called aerosols—usually less than 0.1 μm in diameter); the smallest are gaseous clusters and ions and submicroscopic liquids and solids; somewhat larger ones produce the beautiful blue haze in distant vistas; those 2 to 3 times larger are highly effective in scattering light; and the largest consist of such things as rock fragments, salt crystals, and ashy residues from volcanoes, forest fires, or incinerators.

The numbers of which particulates are concentrated in the atmosphere vary greatly—ranging from more than 10,000,000/cubic centimeter to less than 1/L (0.001/cc). Excluding the particles in gases as well as vegetative material, sizes range from 0.005 to 500 microns, a variation in diameter of 100,000 times.

The largest number of airborne particulates is always in the invisible range. These numbers vary from less than 1 liter to more than a half million per cubic centimeter in heavily polluted air and to at least 10 times more than that when a gas-to-particle reaction is occurring (Schaefer & Day, 1981).

Based on particulate level, there are two distinct regions in the atmosphere: very clean and dirty. In the clean parts, there are so few particulates that they are almost invisible, making them hard to collect or measure. In the dirty parts of the atmosphere—the air of a large metropolitan area—the concentration of particles includes an incredible variety of particulates from a wide variety of sources.

Atmospheric particulate matter performs a number of functions, undergoes several processes, and is involved in many chemical reactions in the atmosphere. Probably the most important function of particulate matter in the atmosphere is its action as nuclei for the formation of water droplets and ice crystals. Much of the work of Vincent J. Schaefer (inventor of cloud seeding) involved using dry ice in early attempts but later evolved around the addition of condensing particles to atmospheres supersaturated with water vapor and the use of silver iodide, which forms huge numbers of very small particles. Another important function of atmospheric particulate matter is that it helps determine the heat balance of the Earth's atmosphere by reflecting light. Particulate matter is also involved in many chemical reactions in the atmosphere such as neutralization, catalytic effects, and oxidation reactions. These chemical reactions will be discussed in greater detail later.

The atmosphere also contains clouds and aerosols (NASA's Earth Observatory library).

Atmosphere lifetime: the average time that a molecule resides in the atmosphere before it is removed by chemical reaction or deposition. In general, if a quantity of a compound is emitted into the atmosphere at a particular time, about 35 percent of that quantity will remain in the atmosphere at the end of the compound's atmospheric lifetime. This fraction will continue to decrease in an exponential way so that about 15 percent of the quantity will remain at the end of two times the atmospheric lifetime, and so on. Greenhouse gas lifetimes can range from a few years to a few thousand years.

Atmospheric deposition: the transfer of substances from the air to the surface of the Earth, either in wet form (rain, fog, snow, dew, frost, hail) or in dry form (gases, aerosols, particles).

Atmospheric pressure: the pressure exerted by the atmosphere on any surface beneath or within it; equal to 14.7 pounds per square inch at sea level.

Average discharge: as used by the U.S. Geological Survey, the arithmetic average of all complete water years of record of surface water discharge whether consecutive or not. The term "average" generally is reserved for the average of record, and "mean" is used for averages of shorter periods, namely daily, monthly, or annual mean discharges.

B

Background concentration: a concentration of a substance in a particular environment that is indicative of minimal influence by human (anthropogenic) sources.

Backwater: a body of water in which the flow is slowed or turned back by an obstruction such as a bridge or dam, an opposing current, or the movement of the tide.

Bacteria: single-celled microscopic organisms.

Bank: the sloping ground that borders a stream and confines the water in the natural channel when the water level, or flow, is normal.

Bank storage: the change in the amount of water stored in an aquifer adjacent to a surface-water body resulting from a change in stage of the surface-water body.

Barrier bar: an elongate offshore ridge, submerged at least at high tide, built up by the action of waves or currents.

Base flow: the sustained low flow of a stream, usually groundwater inflow to the stream channel.

Basic: the opposite of acidic; water that has a pH of greater than 7.

Basin and range physiography: a region characterized by a series of generally north-trending mountain ranges separated by alluvial valleys.

Bedload: sediment that moves on or near the streambed and is in almost continuous contact with the bed.

Bed material: sediment making up the streambed.

Bedrock: a general term used for solid rock that underlies soils or other unconsolidated material.

Bed sediment: the material that is temporarily stationary in the bottom of a stream or other watercourse.

Benthic invertebrates: insects, mollusk, crustaceans, worms, and other organisms without a backbone that live in, on, or near the bottom of lakes, streams, or oceans.

Benthic organism: a form of aquatic life that lives on or near the bottom of stream, lakes, or oceans.

Bioaccumulation: the biological sequestering of a substance at higher concentrations than that at which it occurs in the surrounding environment or medium. Also, the process whereby a substance enters organisms through the gills, epithelial tissues or dietary or other sources.

Bioavailability: the capacity of a chemical constituent to be taken up by living organisms either through physical contact or by ingestion.

Biochemical: refers to chemical processes that occur inside or are mediated by living organisms.

Biochemical oxygen demand (BOD): the amount of oxygen required by bacteria to stabilize decomposable organic matter under aerobic conditions.

Biochemical process: a process characterized by, produced by, or involving chemical reactions in a living organism.

Biodegradation: transformation of a substance into new compounds through biochemical reactions or the actions of microorganisms such as bacteria.

Biodiversity: variety of plant and animal life in the world or in a particular habitat or ecosystem.

Biological treatment: a process that uses living organisms to bring about chemical changes.

Biomass: the amount of living matter, in the form of organisms, present in a particular habitat, usually expressed as weight-per-unit area.

Biota: all living organisms of an area.

Blue hole: a subsurface void, usually a solution sinkhole, developed in carbonate rocks that are open to the Earth's surface and contains tidally influenced waters of fresh, marine, or mixed chemistry.

Bog: a nutrient-poor, acidic wetland dominated by a waterlogged, spongy mat of sphagnum moss that ultimately forms a thick layer of acidic peat; generally has no inflow or outflow; fed primarily by rainwater.

Brackish water: water with a salinity intermediate between seawater and freshwater (containing from 1,000 to 10,000 milligrams per liter of dissolved solids).

Breakdown product: a compound derived by chemical, biological, or physical action upon a pesticide. The breakdown is a natural process that may result in a more or less toxic compound and a more or less persistent compound.

Breakpoint chlorination: the addition of chlorine to water until the chlorine demand has been satisfied and free chlorine residual is available for disinfection.

C

Calcareous: a rock or substance formed of calcium carbonate or magnesium carbonate by biological deposition or inorganic precipitation or containing those minerals in sufficient quantities to effervesce when treated with cold hydrochloric acid.

Capillary fringe: the zone above the water table in which water is held by surface tension. Water in the capillary fringe is under a pressure less than atmospheric.

Carbonate rocks: rocks (such as limestone or dolostone) that are composed primarily of minerals (such as calcite and dolomite) containing a carbonate ion.

Cenote: steep-walled natural well that extends below the water table; generally caused by collapse of a cave roof; term reserved for features found in the Yucatan Peninsula of Mexico.

Center pivot irrigation: an automated sprinkler system involving a rotating pipe or boom that supplies water to a circular area of an agricultural field through sprinkler heads or nozzles.

Channel scour: erosion by flowing water and sediment on a stream channel; results in removal of mud, silt, and sand on the outside curve of a stream bend and the bed material of a stream channel.

Channelization: the straightening and deepening of a stream channel to permit the water to move faster or to drain a wet area for farming.

Chemical treatment: a process that results in the formation of a new substance or substances. The most common chemical water treatment processes include coagulation, disinfection, water softening, and filtration.

Chlordane: octachlor-4,7-methanotetrahydroindane. An organochlorine insecticide no longer registered for use in the U.S. Technically, chlordane is a mixture in which the primary components are cis- and trans-chlordane, cis- and trans-nonachlor, and heptachlor.

Chlorinated solvent: a volatile organic compound containing chlorine. Some common solvents are trichloroethylene, tetrachloroethylene, and carbon tetrachloride.

Chlorination: the process of adding chlorine to water to kill disease-causing organisms or to act as an oxidizing agent.

Chlorine demand: a measure of the amount of chlorine that will combine with impurities and is therefore unavailable to act as a disinfectant.

Chlorofluorocarbons: a class of volatile compounds consisting of carbon, chlorine, and fluorine. Commonly called freons, these have been used in refrigeration mechanisms, as blowing agents in the fabrication of flexible and rigid foams, and, until banned from use several years ago, as propellants in spray cans.

Cienaga: a marshy area where the ground is wet due to the presence of seepage of springs.

Clean Water Act (CWA): federal law dating to 1972 (with several amendments) with the objective to restore and maintain the chemical, physical, and biological integrity of the nation's waters. Its long-range goal is to eliminate the discharge of pollutants into navigable waters and to make national waters fishable and swimmable.

Climate: the sum total of the meteorological elements that characterize the average and extreme conditions of the atmosphere over an extended period of time at any one place or region of the Earth's surface.

Climate change: a change in the state of the climate that can be identified (for example, using statistical tests) by changes in the mean and/or the variability of its properties and that persists for an extended period, typically decades or longer. It refers to any change in climate over time, whether due to natural variability or as result of human activity.

Climate change–related migration: (shorthand: internal climate migration) migration that can be attributed largely to the slow-onset impacts of climate change on livelihoods owing to shifts in water availability and crop productivity or to factors such as sea level rise or storm surge.

Climate migrant/migration: climate migrants are people who move within or between countries because of climate change–related migration.

Coagulants: chemicals that cause small particles to stick together to form larger particles.

Coagulation: a chemical water treatment method that causes exceedingly small, suspended particles to attract one another and form larger particles. This is accomplished by the addition of a coagulant that neutralizes the electrostatic charges that cause particles to repel each other.

Coastal erosion: erosion of coastal landforms that results from wave action, exacerbated by storm surge and sea level rise.

Coliform bacteria: a group of bacteria predominantly inhabiting the intestines of humans or animals but also occasionally found elsewhere. Presence of the bacteria in water is used as an indication of fecal contamination (contamination by animal or human wastes).

Color: a physical characteristic of water. Color is most commonly tan or brown from oxidized iron, but contaminants may cause other colors, such as green or blue. Color differs from turbidity, which is water's cloudiness.

Combined sewer overflow: a discharge of untreated sewage and stormwater to a stream when the capacity of a combined storm/sanitary sewer system is exceeded by storm runoff.

Communicable diseases: usually caused by *microbes*—microscopic organisms, including bacteria, protozoa, and viruses. Most microbes are essential components of our environment and do not cause disease. Those that do are called pathogenic organisms, or simply *pathogens.*

Community: in ecology, the species that interact in a communal area.

Community water system: a public water system that serves at least 15 service connections used by year-round residents or regularly serves at least 25 year-round residents.

Compaction: in this book, compaction is used in its geologic sense and refers to the inelastic compression of the aquifer system. Compaction of the aquifer system reflects the rearrangement of the mineral grain pore structure and largely nonrecoverable reduction of the porosity under stress greater than the preconsolidation stress. Compaction, as used here, is synonymous with the term "virgin consolidation" used by soils engineers. The term refers to both the process and the measured change in thickness. As a practical matter, an exceedingly small amount (1 to 5 percent) of the compaction is recoverable as a slight elastic rebound of the compacted material if stresses are reduced.

Compaction, residual: compaction that would ultimately occur if a given increase in applied stress were maintained until steady-state pore pressures were achieved. Residual compaction may also be defined as the difference between (1) the amount of compaction that will occur ultimately for a given increase in applied stress and (2) that which has occurred at a specified time.

Composite sample: a series of individual or grab samples taken at various times from the same sampling point and mixed together.

Compression: in this book, compression refers to the decrease in thickness of sediments, as a result of increase in vertical compressive stress. Compression many be elastic (fully recoverable) or inelastic (nonrecoverable).

Concentration: the ratio of the quantity of any substance present in a sample of a given volume or a given weight compared to the volume or weight of the sample.

Cone of depression: the depression of heads around a pumping well caused by withdrawal of water.

Confined aquifer (artesian aquifer): an aquifer that is completely filled with water under pressure and that is overlain by material that restricts the movement of water.

Confining bed: a layer of rock having exceptionally low hydraulic conductivity that hampers the movement of water into and out of an aquifer.

Confining layer: a body of impermeable or distinctly less permeable material stratigraphically adjacent to one or more aquifers that restricts the movement of water into and out of the aquifers.

Confining unit: a saturated, relatively low-permeability geologic unit that is areally extensive and serves to confine an adjacent artesian aquifer of aquifers. Leaky confining units may transmit appreciable water to and from adjacent aquifers.

Confluence: the flowing together of two or more streams; the place where a tributary joins the mainstream.

Conglomerate: a coarse-grained sedimentary rock composed of fragments larger than 2 millimeters in diameter.

Consolidation: in soil mechanics, consolidation is the adjustment of a saturated soil in response to increased load, involving the squeezing of water from the pores and decrease in void ratio or porosity of the soul. In this book, the geologic term "compaction" is used in preference to consolidation.

Constituent: a chemical or biological substance in water, sediment, or biota that can be measured by an analytical method.

Consumptiveuse: the quantity of water that is not available for immediate rescue because it has been evaporated, transpired, or incorporated into products, plant tissue, or animal tissue.

Contact recreation: recreational activities, such as swimming and kayaking, in which contact with water is prolonged or intimate and in which there is a likelihood of ingesting water.

Contaminant: a toxic material found as an unwanted residue in or on a substance.

Contamination: degradation of water quality compared to original or natural conditions due to human activity.

Contributing area: the area in a drainage basin that contributes water to streamflow or recharge to an aquifer.

Core sample: a sample of rock, soil, or other material obtained by driving a hollow tube into the undisturbed medium and withdrawing it with its contained sample.

Criterion: a standard rule or test on which a judgment or decision can be based.

Cross connection: any connection between safe drinking water and a nonpotable water or fluid.

CxT value: the product of the residual disinfectant concentration **C**, in milligrams per liter, and the corresponding disinfectant contact time **T**, in minutes. Minimum **CxT** values are specified by the Surface Water Treatment Rule as a means of ensuring adequate kill or inactivation of pathogenic microorganisms in water.

D

Datum plane: a horizontal plane to which ground elevations or water surface elevations are referenced.

Deepwater habitat: permanently flooded lands lying below the deepwater boundary of wetlands.

Deforestation: conversion of forest to non-forest.

Degradation products: compounds resulting from transformation of an organic substance through chemical, photochemical, and/or biochemical reactions.

Denitrification: a process by which oxidized forms of nitrogen such as nitrate are reduced to form nitrites, nitrogen oxides, ammonia, or free nitrogen commonly brought about by the action of denitrifying bacteria and usually resulting in the escape of nitrogen to the air.

Desertification: land degradation in arid, semi-arid, and dry sub-humid areas collectively known as drylands, resulting from many factors, including human activities and climatic variations. The range and intensity of desertification have increased in some dryland areas over the past several decades.

Detection limit: the concentration of a constituent or analyte below which a particular analytical method cannot determine, with a high degree of certainty, the concentration.

Diatoms: single-celled, colonial, or filamentous algae with siliceous cell walls constructed of two overlapping parts.

Direct runoff: the runoff entering stream channels promptly after rainfall or snowmelt.

Discharge: the volume of fluid passing a point per unit of time, commonly expressed in cubic feet per second, million gallons per day, gallons per minute, or seconds per minute per day.

Discharge area (groundwater): area where subsurface water is discharged to the land surface, to surface water, or to the atmosphere.

Disinfectants-disinfection by-products (D-DBPs): a term used in connection with state and federal regulations designed to protect public health by limiting the concentration of either disinfectants or the by-products formed by the reaction of disinfectants with other substances in the water (such as trihalomethanes—THMs).

Disinfection: a chemical treatment method. The addition of a substance (e.g., chlorine, ozone, or hydrogen peroxide) that destroys or inactivates harmful microorganisms or inhibits their activity.

Dispersion: the extent to which a liquid substance introduced into a groundwater system spreads as it moves through the system.

Displacement: forced removal of people or people obliged to flee from their places of habitual residence.

Dissociate: the process of ion separation that occurs when an ionic solid is dissolved in water.

Dissolved constituent: operationally defined as a constituent that passes through a 0.45-micrometer filter.

Dissolved oxygen (DO): the oxygen dissolved in water, usually expressed in milligrams per liter, parts per million, or percent of saturation.

Dissolved solids: any material that can dissolve in water and be recovered by evaporating the water after filtering the suspended material.

Diversion: a turning aside or alteration of the natural course of a flow of water, normally considered physically to leave the natural channel. In some states, this can be a consumptive use directly from another stream, such as by livestock watering. In other states, a diversion must consist of such actions as taking water through a canal, pipe, or conduit.

Dolomite: a sedimentary rock consisting chiefly of magnesium carbonate.

Domestic withdrawals: water used for normal household purposes, such as drinking, food preparation, bathing, washing clothes and dishes, flushing toilets, and watering lawns and gardens. The water may be obtained from a public supplier or may be self-supplied. Also called residential water use.

Drainage area: the drainage area of a stream at a specified location is that area, measured in a horizontal plane, that is enclosed by a drainage divide.

Drainage basin: the land area drained by a river or stream.

Drainage divide: boundary between adjoining drainage basins.

Drawdown: the difference between the water level in a well before pumping and the water level in the well during pumping. Also, for flowing wells, the reduction of the pressure head as a result of the discharge of water.

Drinking water standards: water quality standards measured in terms of suspended solids, unpleasant taste, and microbes harmful to human health. Drinking water standards are included in state water quality rules.

Drinking water supply: any raw or finished water source that is or may be used as a public water system or as drinking water by one or more individuals.

Drip irrigation: an irrigation system in which water is applied directly to the root zone of plants by means of applicators (orifices, emitters, porous tubing, or perforate pipe) operated under low pressure. The applicators can be placed on or below the surface of the ground or can be suspended from supports.

Drought: a prolonged period of less-than-normal precipitation such that the lack of water causes a serious hydrologic imbalance.

E

Ecoregion: an area of similar climate, landform, soil, potential natural vegetation, hydrology, or other ecologically relevant variables.

Ecosystem: a community of organisms considered together with the nonliving factors of the environment.

Effluent: outflow from a particular source, such as a stream that flows from a lake or liquid waste that flows from a factory or sewage-treatment plant.

Effluent limitations: standards developed by the EPA to define the levels of pollutants that could be discharged into surface waters.

Electrodialysis: the process of separating substances in a solution by dialysis, using an electric field as the driving force.

Electronegativity: the tendency for atoms that do not have a complete octet of electrons in their outer shell to become negatively charged.

Ellipsoid, Earth: a mathematically determined three-dimensional surface obtained by rotating an ellipse about its semi-minor axis. In the case of the Earth, the ellipsoid is the modeled shape of its surface, which is relatively flattened in the polar axis.

Ellipsoid, height: the distance of a point above the ellipsoid measured perpendicular to the surface of the ellipsoid.

Emergent plants: erect, rooted, herbaceous plants that may be temporarily or permanently flooded at the base but do not tolerate prolonged inundation of the entire plant.

Enhanced Surface Water Treatment Rule (ESWTR): a revision of the original Surface Water Treatment Rule that includes modern technology and requirements to deal with newly identified problems.

Environment: the sum of all conditions and influences affecting the life of organisms.

Environmental mobility: temporary or permanent mobility as a result of sudden or progressive changes in the environmental that adversely affect living conditions, either within countries or across borders.

Environmental sample: a water sample collected from an aquifer or stream for the purpose of chemical, physical, or biological characterization of the sampled resource.

Environmental setting: land area characterized by a unique combination of natural and human-related factors, such as row-crop cultivation or glacial-till soils.

Ephemeral stream: a stream or part of a stream that flows only in direct response to precipitation; it receives little or no water from springs, melting snow, or other sources; its channel is at all times above the water table.

EPT richness index: an index based on the sum of the number of taxa in three insect orders, Ephemeroptera (mayflies), Plecoptera (stoneflies), and Trichoptera (caddisflies), that are composed primarily of species considered relatively intolerant to environmental alterations.

Equipotential line: a line on a map or cross-section along which total heads are the same.

Erosion: the process whereby materials of the Earth's crust are loosened, dissolved, or worn away and simultaneously moved from one place to another.

Eutrophication: the process by which water becomes enriched with plant nutrients, most commonly phosphorus and nitrogen.

Evacuation: moving people and assets temporarily to safer places before, during, or after the occurrence of a hazardous event in order to protect them.

Evaporite minerals (deposits): minerals or deposits of minerals formed by evaporation of water containing salts. These deposits are common in arid climates.

Evaporites: a class of sedimentary rocks composed primarily of minerals precipitated from a saline solution as a result of extensive or total evaporation of water.

Evapotranspiration: the process by which water is discharged to the atmosphere as a result of evaporation from the soil and surface-water bodies and transpiration by plants.

Exfoliation: the process by which concentric scales, plates, or shells or rock, from less than centimeter to several meters in thickness, are stripped from the bare surface of a large rock mass.

Extreme heat event: three or more days of above-average temperatures, generally defined as passing a certain threshold (for example, above the 85th percentile for average daily temperature in a year).

Extreme weather event: weather event that is rare at a particular place and time of year with characteristics of extreme weather varying from place to place in an absolute sense. When a pattern of extreme weather persists for some time, such as a season, it may be classified as an extreme climate event, especially if it yields an average or total that is itself extreme (for example, drought or heavy rainfall over a season).

F

Facultative bacteria: a type of anaerobic bacteria that can metabolize its food either aerobically or anaerobically.

Fall line: imaginary line marking the boundary between the ancient, resistant crystalline rocks of the Piedmont province of the Appalachian Mountains and the younger, softer sediments of the Atlantic Coastal Plain province in the eastern United States. Along rivers, this line commonly is reflected by waterfalls.

Fecal bacteria: microscopic single-celled organisms (primarily fecal coliforms and fecal streptococci) found in the wastes of warm-blooded animals. Their presence in water is used to assess the sanitary quality of water for body-contact recreation or for consumption. Their presence indicates contamination by the wastes of warm-blooded animals and the possible present of pathogenic (disease-producing) organisms.

Federal Water Pollution Control Act (1972): The act outlines the objective "to restore and maintain the chemical, physical, and biological integrity of the nation's waters." This 1972 act and subsequent Clean Water Act amendments are the most far-reaching water pollution control legislation ever enacted. They provided for comprehensive programs for water pollution control, uniform laws, and interstate cooperation. They provided grants for research, investigations, training, and information on national programs on surveillance, the effects of pollutants, pollution control, and the identification and measurement of pollutants. Additionally, they allot grants and loans for the construction of treatment works. The act established national discharge standards with enforcement provisions.

The Federal Water Pollution Control Act established several milestone achievement dates. It required secondary treatment of domestic waste by publicly owned treatment works (POTWs) and application of "best practicable" water pollution control technology by industry by 1977. Virtually all industrial sources have achieved compliance (because of economic difficulties and cumbersome federal requirements, certain POTWs obtained an extension to July 1, 1988, for compliance). The act also called for new levels of technology to be imposed during the 1980s and 1990s, particularly for controlling toxic pollutants.

The act mandates a strong pretreatment program to control toxic pollutants discharged by industry into POTWs. The 1987 amendments require that stormwater from industrial activity must be regulated.

Fertilizer: any of a large number of natural or synthetic materials, including manure and nitrogen, phosphorus, and potassium compound, spread on or worked into soil to increase its fertility.

Filtrate: liquid that has been passed through a filter.

Filtration: a physical treatment method for removing solid (particulate) matter from water by passing the water through porous media such as sand or a human-made filter.

Flocculation: the water treatment process following coagulation, it uses gentle stirring to bring suspended particles together so that they will form larger, more settleable clumps called floc.

Flood: any relatively high streamflow that overflows the natural or artificial banks of a stream.

Flood attenuation: a weakening or reduction in the force or intensity of a flood.

Flood irrigation: the application of irrigation water whereby the entire surface of the soil is covered by ponded water.

Flood plain: a strip of relatively flat land bordering a stream channel that is inundated at times of high water.

Flow line: the idealized path followed by particles of water.

Flow net: The grid pattern formed by a network of flow lines and equipotential lines.

Flowpath: an underground route for ground-water movement, extending from a recharge (intake) zone to a discharge (output) zone such as a shallow stream.

Fluvial: pertaining to a river or stream.

Forced migration: migratory movement in which an element of coercion exists, including threat to life and livelihood, whether arising from natural or human-made causes. This includes movements of refuges and internally displaced persons as well as people displaced by natural or environmental disasters, chemical or nuclear disasters, famine, or development projects.

Freshwater: water that contains less than 1,000 milligrams per liter of dissolved solids.

Freshwater chronic criteria: the highest concentration of a contaminant that freshwater aquatic organisms can be exposed to for an extended period of time (four days) without adverse effects.

G

Geodetic datum: a set of constants specifying the coordinate system used for geodetic control, for example, for calculating the coordinates of points on the Earth.

Geoid, Earth: The sea-level equipotential surface or figure of the Earth. If the Earth were completely covered by a shallow sea, the surface of this sea would conform to the geoid shaped by the hydrodynamic equilibrium of the water subject to gravitational and rotational forces. Mountains and valleys are departures from the reference geoid.

Grab sample: a single water sample collected at one time from a single point.

Groundwater: the fresh water found under the Earth's surface, usually in aquifers. Groundwater is a major source of drinking water and a source of a growing concern in areas where leaching agricultural or industrial pollutants or substances from leaking underground storage tanks are contaminating groundwater.

H

Habitat: the part of the physical environment in which a plant or animal lives.

Hardness: a characteristic of water caused primarily by the salts of calcium and magnesium. It causes deposition of scale in boilers, damage in some industrial processes, and sometimes objectionable taste. It may also decrease soap's effectiveness.

Hazard: the potential occurrence of a natural or human-induced physical event or trend or physical impact that may cause loss of life, injury, or other health impacts, as well as damage and loss to property, infrastructure, livelihoods, service provision, ecosystems, and environmental resources.

Head: the height above a datum plane of a column of water. In a groundwater system, it is composed of elevation head and pressure head.

Headwaters: the source and upper part of a stream.

Hydraulic conductivity: the capacity of a rock to transmit water. It is expressed as the volume of water at the existing kinematic viscosity that will move in unit time under a unit hydraulic gradient through a unit area measured at right angles to the direction of flow.

Hydraulic gradient: the change of hydraulic head per unit of distance in a given direction.

Hydrocompaction: the process of volume decrease and density increase that occurs when certain moisture-deficient deposits compact as they are wetted for the first time since burial. The vertical downward movement of the land surface that results from this process has also been termed "shallow subsidence" and "near-surface subsidence."

Hydrogen bonding: the term used to describe the weak but effective attraction that occurs between polar covalent molecules.

Hydrograph: graph showing variation of water elevation, velocity, streamflow, or another property of water with respect to time.

Hydrologic cycle: literally the water-earth cycle. The movement of water in all three physical forms through the various environmental mediums (air, water, biota, and soil).

Hydrology: the science that deals with water as it occurs in the atmosphere, on the surface of the ground, and underground.

Hydrostatic pressure: the pressure exerted by the water at any given point in a body of water at rest.

Hygroscopic: a substance that readily absorbs moisture.

I

Immobility: inability to move from a place of risk or not moving away from a place of risk due to choice.

Impermeability: the incapacity of a rock to transmit a fluid.

Index of biotic integrity (IBI): an aggregated number, or index, based on several attributes or metrics of a fish community that provides an assessment of biological conditions.

Indicator sites: stream sampling sites located at outlets of drainage basins with relatively homogeneous land use and physiographic conditions; most indicator-site basins have drainage areas ranging from 20 to 200 square miles.

Infiltration: the downward movement of water from the atmosphere into soil or porous rock.

Influent: water flowing into a reservoir, basin, or treatment plant.

Inorganic: containing no carbon; matter other than plant or animal.

Inorganic chemical: a chemical substance of mineral origin not having carbon in its molecular structure.

Inorganic soil: soil with less than 20 percent organic matter in the upper 16 inches.

Ionic bond: the attractive forces between oppositely charged ions—for example, the forces between the sodium and chloride ions in a sodium chloride crystal.

Instantaneous discharge: the volume of water that passes a point at a particular instant of time.

Instream use: water use taking place within the stream channel for such purposes as hydroelectric power generation, navigation, water-quality improvement, fish propagation, and recreation. Sometimes called nonwithdrawal use or in-channel use.

Intermittent stream: a stream that flows only when it receives water from rainfall runoff or springs or from some surface source such as melting snow.

Internal drainage: surface drainage whereby the water does not reach the ocean, such as drainage toward the lowermost or central part of an interior basin or closed depression.

Internal migration (migrant): migration within national borders.

International migration (migrant): migration that occurs across national borders.

Intertidal: alternately flooded and exposed by tides.

Intolerant organisms: organisms that are not adaptable to human alterations to the environment and thus decline in numbers where alterations occur.

Invertebrate: an animal having no backbone or spinal column.

Ion: a positively or negatively charged atom or group of atoms.

Irregular migration: movement of persons that takes place outside the laws, regulations, or international agreements governing the entry into or exit from the state of origin, transit, or destination.

Irrigation: controlled application of water to arable land to supply requirements of crops not satisfied by rainfall.

Irrigation return flow: the part of irrigation applied to the surface that is not consumed by evapotranspiration or uptake by plants and that migrates to an aquifer or surface-water body.

Irrigation withdrawals: withdrawals of water for application on land to assist in the growing of crops and pastures or to maintain recreational lands.

K

Karst: a type of topography that is formed on limestone, dolomite, gypsum, and other rocks, primarily by dissolution, and that is characterized by sinkholes, caves, and subterranean drainage.

Karstification: action by water, mainly chemical but also mechanical, that produces features of a karst topography.

Karst mantled: a terrane of karst features, usually subdued, and covered by soil or a thin alluvium.

Karst topography: type of topography that is formed in limestone, gypsum, and other similar-type rock by dissolution and is characterized by sinkholes, caves, and rapid underground water movement.

Kill: Dutch term for stream or creek.

L

Lacustrine: pertaining to, produced by, or formed in a lake.

Leachate: a liquid that has percolated through soil containing soluble substances and that contains certain amounts of these substances in solution.

Leaching: the removal of materials in solution from soil or rock; also refers to movement of pesticides or nutrients from land surface to groundwater.

Limnetic: the deepwater zone (greater than 2 meters deep).

Littoral: the shallow-water zone (less than 2 meters deep).

Load: material that is moved or carried by streams, reported as weight of material transported during a specified time period, such as tons per year.

M

Main stem: the principal trunk of a river or a stream.

Marsh: a water-saturated, poorly drained area, intermittently or permanently water covered, having aquatic and grasslike vegetation.

Maturity (stream): the stage in the development of a stream at which it has reached its maximum efficiency, when velocity is just sufficient to carry the sediment delivered to it by tributaries; characterized by a broad, open, flat-floored valley having a moderate gradient and gentle slope.

Maximum contaminant level (MCL): a primary standard, whereas an MCLG is a maximum concentration goal for a drinking water contaminant, which would be desirable based on human health concerns and assuming all feasibility issues such as cost and technological capacity are not considered. Stated differently, MCL is the maximum allowable concentration of a contaminant in drinking water, as established by state and/or federal regulations. Primary MCLs are health related and mandatory. Secondary MCLs are related to the aesthetics of the water and are highly recommended but not required.

Mean discharge: the arithmetic mean of individual daily mean discharges of a stream during a specific period, usually daily, monthly, or annually.

Membrane filter method: a laboratory method used for coliform testing. The procedure uses an ultra-thin filter with a uniform pore size smaller than bacteria (less than a micron). After water is forced through the filter, the filter is incubated in a special media that promotes the growth of coliform bacteria. Bacterial colonies with a green-gold sheen indicate the presence of coliform bacteria.

Method detection limit: the minimum concentration of a substance that can be accurately identified and measured with current lab technologies.

Midge: a small fly in the family Chironomidae. The larval (juvenile) life stages are aquatic.

Minimum reporting level (MRL): the smallest measured concentration of a constituent that may be reliably reported using a given analytical method. In many cases, the MRL is used when documentation for the method detection limit is not available.

Mitigation: actions taken to avoid, reduce, or compensate for the effects of human-induced environmental damage.

Modes of transmission of disease: the ways in which diseases spread from one person to another.

Monitoring: repeated observation, measurement, or sampling at a site, on a scheduled or event basis, for a particular purpose.

Monitoring well: a well designed for measuring water levels and testing groundwater quality.

N

National Primary Drinking Water Regulations (NPDWRs): regulations developed under the Safe Drinking Water Act, which establish maximum contaminant levels, monitoring requirements, and reporting procedures for contaminants in drinking water that endanger human health.

National Pollutant Discharge Elimination System (NPDES): a requirement of the CWA that discharges meet certain requirements prior to discharging waste to any water body. It sets the highest permissible effluent limits, by permit, prior to making any discharge.

Near Coastal Water Initiative: this initiative was developed in 1985 to provide for management of specific problems in waters near coastlines that are not dealt with in other programs.

Nitrate: an ion consisting of nitrogen and oxygen (NO_3). Nitrate is a plant nutrient and is very mobile in soils.

Nonbiodegradable: substances that do not break down easily in the environment.

Nonpoint source: a source (of any water-carried material) from a broad area, rather than from discrete points.

Nonpoint-source contaminant: a substance that pollutes or degrades water that comes from lawn or cropland runoff, the atmosphere, roadways, and other diffuse sources.

Nonpoint-source water pollution: water contamination that originates from a broad area (such as leaching of agricultural chemicals from crop lad) and enters the water resource diffusely over a large area.

Nonpolar covalently bonded: a molecule composed of atoms that share their electrons equally, resulting in a molecule that does not have polarity.

Nutrient: any inorganic or organic compound needed to sustain plant life.

O

Organic: containing carbon, but possibly also containing hydrogen, oxygen, chlorine, nitrogen, and other elements.

Organic chemical: a chemical substance of animal or vegetable origin having carbon in its molecular structure.

Organic detritus: any loose organic material in streams—such as leaves, bark, or twigs—removed and transported by mechanical means, such as disintegration or abrasion.

Organic soil: soil that contains more than 20 percent organic matter in the upper 16 inches.

Organochlorine compound: synthetic organic compounds containing chlorine. As generally used, the term refers to compounds containing mostly or exclusively carbon, hydrogen, and chlorine.

Outwash: soil material washed down a hillside by rainwater and deposited upon more gently sloping land.

Overdraft: any withdrawal of groundwater in excess of the *safe yield*.

Overland flow: the flow of rainwater or snowbelt over the land surface toward stream channels.

Oxidation: when a substance either gains oxygen or loses hydrogen or electrons in a chemical reaction. One of the chemical treatment methods.

Oxidizer: a substance that oxidizes another substance.

P

Paleokarst: a karstified area that has been buried by later deposition of sediments.

Parts per million: the number of weight or volume units of a constituent present with each one million units of the solution or mixture. Formerly used to express the results of most water and wastewater analyses, PPM is being replaced by milligrams per liter, M/L. For drinking water analyses, concentration in parts per million and milligrams per liter are equivalent. A single PPM can be compared to a shot glass full of water inside a swimming pool.

Pathogens: types of microorganisms that can cause disease.

Perched groundwater: unconfined groundwater separated from an underlying main body of groundwater by an unsaturated zone.

Percolation: the movement, under hydrostatic pressure, of water through interstices of a rock or soil (except the movement through large openings such as caves).

Perennial stream: a stream that normally has water in its channel at all times.

Periphyton: microorganisms that coat rocks, plants, and other surfaces on lake bottoms.

Permeability: the capacity of a rock for transmitting a fluid; a measure of the relative ease with which a porous medium can transmit a liquid. The quality of the soil that enables water to move downward through the soil profile. Permeability is measured as the number of inches per hour that water moves downward through the saturated soil. Terms describing permeability are:

Very slow	less than 0.06 inches/hr
Slow	0.06 to 0.2 inches/hr
Moderately slow	0.2 to 0.6 inches/hr
Moderate	0.6 to 2.0 inches/hr
Moderately rapid	2.0 to 6.0 inches/hr
Rapid	6.0 to 20 inches//hr
Very rapid	more than 20 inches/hr

pH: a measure of the acidity (less than 7) or alkalinity (greater than 7) of a solution; a pH of 7 is considered neutral.

Phosphorus: a nutrient essential for growth that can play a key role in stimulating aquatic growth in lakes and streams.

Photosynthesis: the synthesis of compounds with the aid of light.

Physical treatment: any process that does not produce a new substance (e.g., screening, adsorption, aeration, sedimentation, and filtration).

Planned relocation: people moved or assisted to move permanently away from areas of environmental risks.

Plutonic: a loosely defined term with a number of current usages. We use it to describe igneous rock bodies that crystallized at great depth or, more generally, any intrusive igneous rock.

Point source: originating at any discrete source.

Polar covalent bond: the shared pair of electrons between two atoms are not equally held. Thus, one of the atoms becomes slightly positively charged, and the other atom becomes slightly negatively charged.

Polar covalent molecule: (water) one or more polar covalent bonds result in a molecule that is polar covalent. Polar covalent molecules exhibit partial positive and negative poles, causing them to behave like tiny magnets. Water is the most common polar covalent substance.

Pollutant: any substance introduced into the environment that adversely affects the usefulness of the resource.

Pollution: the presence of matter or energy whose nature, location, or quantity produces undesired environmental effects. Under the Clean Water Act, for example, the term is defined as a human-made or human-induced alteration of the physical, biological, and radiological integrity of water.

Polychlorinated biphenyls (PCBs): a mixture of chlorinated derivatives of biphenyl, marketed under the trade name Aroclor with a number designating the chlorine content (such as Aroclor 1260). PCBs were used in transformers and capacitors for insulating purposes and in gas pipeline systems as a lubricant. Further sale or new use was banned by law in 1979.

Polycyclic aromatic hydrocarbon (PAH): a class of organic compounds with a fused-ring aromatic structure. PAHs result from incomplete combustion of organic carbon (including wood), municipal solid waste, and fossil fuels, as well as from natural or anthropogenic introduction of uncombusted coal and oil. PAHs included benzo(a)pyrene, fluoranthene, and pyrene.

Population: a collection of individuals of one species or mixed species making up the residents of a prescribed area.

Porosity: 1. The ratio of the aggregate volume of pore spaces in rock or soil to its total volume, usually stated as a percent. 2. A measure of the water-bearing capacity of subsurface rock. With respect to water movement, it is not just the total magnitude of porosity that is important but the size of the voids and the extent to which they are interconnected, as the pores in a formation may be open, interconnected, or closed and isolated. For example, clay may have a very high porosity with respect to potential water content, but it constitutes a poor medium as an aquifer because the pores are usually so small.

Potable water: water that is safe and palatable for human consumption.

Potentiometric surface: a surface that represents the total head in an aquifer; that is, it represents the height above a datum plane at which the water level stands in tightly cased wells that penetrated the aquifer.

Precipitation: any or all forms of water particles that fall from the atmosphere, such as rain, snow, hail, and sleet. The act or process of producing a solid phase within a liquid medium.

Pretreatment: any physical, chemical, or mechanical process used before the main water treatment processes. It can include screening, presedimentation, and chemical addition.

Primary Drinking Water Standards: regulations on drinking water quality (under SWDA) considered essential for preservation of public health.

Primary treatment: the first step of treatment at a municipal wastewater treatment plant. It typically involves screening and sedimentation to remove materials that float or settle.

Public-supply withdrawals: water withdrawn by public and private water suppliers for use within a general community. Water is used for a variety of purposes such as domestic, commercial, industrial, and public water use.

Public water system: as defined by the Safe Drinking Water Act, any system, publicly or privately owned, that serves at least 15 service connections 60 days out of the year or serves an average of 25 people at least 60 days out of the year.

Publicly owned treatment works (POTW): waste treatment works owned by a state, local government unit, or Indian tribe, usually designed to treat domestic wastewaters.

R

Rainfed agriculture: agricultural practice relying almost entirely on rainfall as its source of water. Because of increased weather variability, climate change is expected to make rain-fed farmers more vulnerable to climate change.

Rain shadow: a dry region on the lee side of a topographic obstacle, usually a mountain range, where rainfall is noticeably less than on the windward side.

Rapid-onset event: event such as cyclones and floods which take place in days or weeks (in contrast to slow-onset climate changes that occur over extended periods of time).

Reach: a continuous part of a stream between two specified points.

Reaeration: the replenishment of oxygen in water from which oxygen has been removed.

Receiving waters: a river, lake, ocean, stream, or other water source into which wastewater or treated effluent is discharged.

Recharge: the process by which water is added to a zone of saturation by percolation from the soil surface or by artificial injection.

Recharge area (groundwater): an area within which water infiltrates the ground and reaches the zone of saturation.

Reference dose (RfD): an estimate of the amount of a chemical that a person can be exposed to on a daily basis that is not anticipated to cause adverse systemic health effects over the person's lifetime.

Refoulment: the act of forcing a refugee or asylum seeker to return to a country or territory when he or she is in danger of persecution or torture.

Regular migration: migration that occurs in compliance with the laws of the country of origin, transit, and destination.

Relative sea level rise: land subsidence plus global sea level rise is the contributor to relative sea level rise and, if not abated, will soon (in less than 150 years) inundate many of the major cities and other low-lying areas in the region.

Representative sample: a sample containing all the constituents present in the water from which it was taken.

Return flow: that part of irrigation water that is not consumed by evapotranspiration and that returns to its source or another body of water.

Reverse osmosis (RO): solutions of differing ion concentrations are separated by a semipermeable membrane. Typically, water flows from the chamber with lesser ion concentration into the chamber with the greater ion concentration, resulting in hydrostatic or osmotic pressure. In RO, enough external pressure is applied to overcome this hydrostatic pressure, thus reversing the flow of water. This results in the water on the other side of the membrane becoming depleted in ions and demineralized.

Riffle: a shallow part of the stream where water flows swiftly over completely or partially submerged obstructions to produce surface agitation.

Riparian: pertaining to or situated on the bank of a natural body of flowing water.

Riparian rights: a concept of water law under which authorization to use water in a stream is based on ownership of the land adjacent to the stream.

Riparian zone: pertaining to or located on the bank of a body of water, especially a stream.

Rock: any naturally formed consolidated or unconsolidated material (but not soil) consisting of two or more minerals

Runoff: that part of precipitation or snowmelt that appears in streams or surface-water bodies.

Rural withdrawals: water used in suburban or farm areas for domestic and live-stock needs. The water generally is self supplied and includes domestic use, drinking water for livestock, and other uses such as dairy sanitation, evaporation from stock-watering ponds, and cleaning and waste disposal.

S

Safe Drinking Water Act (SDWA): a federal law passed in 1974 with the goal of establishing federal standards for drinking water quality, protecting underground sources of water, and setting up a system of state and federal cooperation to assure compliance with the law.

Saline water: water that is considered unsuitable for human consumption or for irrigation because of its high content of dissolved solids; generally expressed as milligrams per liter (mg/L) of dissolved solids; seawater is generally considered to contain more than 35,000 mg/L of dissolved solids. A general salinity scale is:

Concentration of Dissolved Solids in mg/L	
Slightly saline	1,000–3,000
Moderately saline	3,000–10,000
Very saline	10,000–35,000
Brine	More than 35,000

Saturated zone: a subsurface zone in which all the interstices or voids are filled with water under pressure greater than that of the atmosphere.

Screening: a pretreatment method that uses coarse screens to remove large debris from the water to prevent clogging of pipes or channels to the treatment plant.

Sea level rise: increases in the height of the sea with respect to a specific point on land.

Secondary Drinking Water Standards: regulations developed under the Safe Drinking Water Act that established maximum levels of substances affecting the aesthetic characteristics (taste, color, or odor) of drinking water.

Secondary maximum contaminant level (SMCL): the maximum level of a contaminant or undesirable constituent in public water systems that, in the judgment of US EPA, is required to protect the public welfare. SMCLs are secondary (nonenforceable) drinking water regulations established by the US EPA for contaminants that may adversely affect the odor or appearance of such water.

Secondary treatment: the second step of treatment at a municipal wastewater treatment plant. This step uses growing numbers of microorganisms to digest organic matter and reduce the amount of organic waste. Water leaving this process is chlorinated to destroy any disease-causing microorganisms before its release.

Sedimentation: a physical treatment method that involves reducing the velocity of water in basins so that the suspended material can settle out by gravity.

Seep: a small area where water percolates slowly to the land surface.

Seiche: a sudden oscillation of the water in a moderate-size body of water, caused by wind.

Sinkhole: a depression in a karst area. At land surface, its shape is general circular and its size measured in meters to tens of meters; underground it is commonly funnel-shaped and associated with subterranean drainage.

Sinuosity: the ratio of the channel length between two points on a channel to the straight-line distance between the same two points; a measure of meandering.

Slow-onset climate change: changes in climate parameters—such as temperature, precipitation, and associated impacts, such as water availability and crop production declines—that occur over extended periods of time.

Soil: the layer of material at the land surface that supports plant growth.

Soil horizon: a layer of soil that is distinguishable from adjacent layers by characteristic physical and chemical properties.

Soil moisture: water occurring in the pore spaces between the soil particles in the unsaturated zone from which water is discharged by the transpiration of plants or by evaporation from the soil.

Solution: formed when a solid, gas, or another liquid in contact with a liquid becomes dispersed homogeneously throughout the liquid. The substance, called a solute, is said to dissolve. The liquid is called the solvent.

Solvated: when either a positive or negative ion becomes completely surrounded by polar solvent molecules.

Sorb: to take up and hold either by absorption or adsorption.

Sorption: general term for the interaction (binding or association) of a solute ion or molecule with a solid.

Spall: a chip or fragment removed from a rock surface by weathering; especially by the process of exfoliation.

Specific capacity: the yield of a well per unit of drawdown.

Specific retention: the ratio of the volume of water retained in a rock after gravity drainage to the volume of the work.

Specific storage: the volume of water that an aquifer system releases or takes into storage per unit volume per unit change in head. The specific storage is equivalent to the *storage coefficient* divided by the thickness of the aquifer system.

Specific yield: the ratio of the volume of water that will drain under the influence of gravity to the volume of saturated rock.

Spring: place where any natural discharge of groundwater flows at the ground surface.

Stateless person: a person who is not considered a national by any state under the operation of its law.

Storage: the capacity of an aquifer, aquitard, or aquifer system to release or accept water into groundwater storage, per unit change in hydraulic head.

Storage coefficient: the volume of water released from storage in a unit prism of an aquifer when the head is lowered a unit distance.

Strain: Deformation that results from a stress. Expressed in terms of the amount of deformation per inch.

Stratification: the layered structure of sedimentary rocks.

Stress and strain: in materials, stress is a measure of the deforming force applied to a body. Strain (which is often erroneously used as a synonym for stress) is really the resulting change in its shape (deformation). For perfectly elastic material, stress is proportional to stain. This relationship is explained by Hooke's law, which states that the deformation of a body is proportional to the magnitude of the deforming force, provided that the body's elastic limit is not exceeded. If the elastic limit is not reached, the body will return to its original size once the force is removed. For example, if a spring is stretched by 2 cm by a weight of 1 N, it will be stretched by 4 cm by a weight of 2 N, and so on; however, once the load exceeds the elastic limit for the spring, Hooke's law will no longer be obeyed, and each successive increase in weight will result in a greater extension until the spring finally breaks. Stress forces are categorized in three ways:

1. Tension (or tensile stress), in which equal and opposite forces that act away from each other are applied to a body; tends to elongate a body.
2. Compression stress, in which equal and opposite forces that act toward each other are applied to a body; tends to shorten it.
3. Shear stress, in which equal and opposite forces that do not act along the same line of action or plane are applied to a body; tends to change its shape without changing its volume.

Stress, geostatic (lithostatic): the total weight (per unit area) of sediments and water above some plane of reference. Geostatic stress normal to any horizontal plane of reference in a saturated deposit may also be defined as the sum of the effective stress and the fluid pressure at that depth.

Stress, preconsolidation: the maximum antecedent effective stress to which a deposit has been subjected and which it can with stand without undergoing additional permanent deformation. Stress changes in the range less than the preconsolidation stress produce elastic deformations of small magnitude. In fine-grained materials, stress increase beyond the preconsolidation stress produced much larger deformations that are principally inelastic (nonrecoverable). Synonymous with "virgin stress."

Stress, seepage: force (per unit area) transferred from the water to the medium by viscous friction when water flows through a porous medium. The force transferred to the medium is equal to the loss of hydraulic head and is termed the seepage force exerted in the friction of flow.

Stressor: event or trend that has an important effect on the system exposed and can increase vulnerability to climate-related risk.

Subsidence: a dropping of the land surface as a result of ground water being pumped. Cracks and fissures can appear in the land. Some state that subsidence is virtually an irreversible process. Others, like the author of this book, state that the jury is still out on the validity of this statement; HRSD's SWIFT project may prove that land subsidence can be reversed.

Subsistence agriculture: occurs when farmers grow food crops to meet the needs of themselves and their families on smallholdings. Subsistence agriculturalists target farm output for survival and for mostly local requirements, with little or no surplus.

Surface runoff: runoff that travels over the land surface to the nearest stream channel.

Surface tension: the attractive forces exerted by the molecules below the surface upon those at the surface, resulting in them crowding together and forming a higher density.

Surface water: all water naturally open to the atmosphere, and all springs, wells, or other collectors that are directly influenced by surface water.

Surface Water Treatment Rule (SWTR): a federal regulation established by the USEPA under the Safe Drinking Water Act that imposes specific monitoring and treatment requirements on all public drinking water systems that draw water from a surface water source.

Suspended sediment: sediment that is transported in suspension by a stream.

Suspended solids: different from suspended sediment only in the way that the sample is collected and analyzed.

Sustainable livelihood: livelihood that endures over time and is resilient to the impacts of several types of shocks, including climate and economic.

Synthetic organic chemicals (SOCs): generally applied to manufactured chemicals that are not as volatile as volatile organic chemicals. Included are herbicides, pesticides, and chemicals widely used in industries.

T

Total head: the height above a datum plane of a column of water. In a groundwater system, it is composed of elevation head and pressure head.

Total suspended solids (TSS): solids present in wastewater.

Transmissivity (groundwater): The capacity of a rock to transmit water under pressure. The coefficient of transmissibility is the rate of flow of water, at the prevailing water temperature, in gallons per day, through a vertical strip of the aquifer 1 foot wide, extending the full saturated height of the aquifer under a hydraulic gradient of 100 percent. A hydraulic gradient of 100 percent means a 1-foot drop in head in 1 foot of flow distance.

Transpiration: the process by which water passes through living organisms, primarily plants, into the atmosphere.

Trihalomethanes (THMs): a group of compounds formed when natural organic compounds from decaying vegetation and soil (such as humic and fulvic acids) react with chlorine.

Turbidity: a measure of the cloudiness of water caused by the presence of suspended matter, which shelters harmful microorganisms and reduces the effectiveness of disinfecting compounds.

U

Unconfined aquifer: an aquifer whose upper surface is a water table free to fluctuate under atmospheric pressure.

Unsaturated zone: a subsurface zone above the water table in which the pore spaces may contain a combination of air and water.

V

Vehicle of disease transmission: any nonliving object or substance contaminated with pathogens.

Vernal pool: a small lake or pond that is filled with water for only a brief time during the spring.

Vug: a small cavity or chamber in rock that may be lined with crystals.

Vulnerability: propensity or predisposition to be adversely affected. Vulnerability encompasses a variety of concepts and elements, including sensitivity or susceptibility to harm and lack of capacity to cope and adapt.

W

Wastewater: the spent or used water from individual homes, a community, a farm, or an industry that contains dissolved or suspended matter.

Waterborne disease: water is a potential vehicle of disease transmission, and waterborne disease is possibly one of the most preventable types of communicable illness. The application of basic sanitary principles and technology has virtually eliminated serious outbreaks of waterborne diseases in developed countries. The most prevalent waterborne diseases include *typhoid fever, dysentery, cholera, infectious hepatitis*, and *gastroenteritis*.

Water budget: an accounting of the inflow to, outflow from, and storage changes of water in a hydrologic unit.

Water column: an imaginary column extending through a water body from its floor to its surface.

Water demand: water requirements for a particular purpose, such as irrigation, power, municipal supply, plant transpiration, or storage.

Water table: the top water surface of an unconfined aquifer at atmospheric pressure.

Water softening: a chemical treatment method that uses either chemicals to precipitate or a zeolite to remove those metal ions (typically $Ca2+$, $Mg2+$, $Fe3+$) responsible for hard water.

Watershed: the land area that drains into a river, river system, or other body of water.

Wellhead protection: the protection of the surface and subsurface areas surrounding a water well or well field supplying a public water system from contamination by human activity.

Y

Yield: the mass of material or constituent transported by a river in a specified period of time divided by the drainage area of the river basin.

Yield, optimal: an optimal amount of groundwater, by virtue of its use, that should be withdrawn from an aquifer system or groundwater basin each year. It is a dynamic quantity that must be determined from a set of alternative groundwater management decisions subject to goals, objectives, and constraints of the management plan.

Yield, perennial: the amount of usable water from an aquifer than can be economically consumed each year for an indefinite period of time. It is a specified amount that is commonly specified equal to the mean annual recharge to the aquifer system, which thereby limits the amount of groundwater that can be pumped for beneficial use.

Yield, safe: the amount of groundwater that can be safely withdrawn from a groundwater basin annually without producing an undesirable result. Undesirable results include but are not limited to depletion of groundwater storage, intrusion of water of undesirable quality, contravention of existing water rights, deterioration of the economic advantages of pumping (such as excessively lower water level sand the attendant increased pumping lifts and associated energy costs), excessive depletion of stream flow by induced infiltration, and land subsidence.

Z

Zone of aeration: the zone above the water table. Water in the zone of aeration does not flow into a well.

Zone of capillarity: the area above a water table where some or all of the interstices (pores) are filled with water that is held by capillarity.

Index

Note: Page numbers in *italics* indicate a figure and page numbers in **bold** indicate a table on corresponding page.

A

For Product Safety Concerns and Information please contact our EU
representative GPSR@taylorandfrancis.com
Taylor & Francis Verlag GmbH, Kaufingerstraße 24, 80331 München, Germany